Lighter than Air Robots

International Series on
INTELLIGENT SYSTEMS, CONTROL, AND AUTOMATION: SCIENCE AND ENGINEERING

VOLUME 58

For further volumes:
www.springer.com/series/6259

Yasmina Bestaoui Sebbane

Lighter than Air Robots

Guidance and Control of Autonomous Airships

Springer

Yasmina Bestaoui Sebbane
Electrical Engineering
Université d'Evry
Rue du Pelvoux 38
Evry 91020
France
Yasmina.Bestaoui@iup.univ-evry.fr

ISBN 978-94-007-2662-8 e-ISBN 978-94-007-2663-5
DOI 10.1007/978-94-007-2663-5
Springer Dordrecht Heidelberg London New York

Library of Congress Control Number: 2011941682

Cover design: VTeX UAB, Lithuania

Printed on acid-free paper

Springer is part of Springer Science+Business Media (www.springer.com)

To my family

Preface

An aerial robot is a system capable of sustained flight with no direct human control and able to perform a specific task. A lighter than air robot can also be defined as a lighter than air unmanned aerial vehicle or an unmanned airship with sufficient autonomy. Lighter than air systems are particularly appealing since the energy to keep them airborne is small. To provide autonomy to the aerial vehicles requires the development of methods to conducting and decisions for implementing the various operations of a mission. Guidance and control for lighter than air robots involve significant differences from traditionally defined mobile robots. Qualities characteristics to lighter than air robots include non trivial dynamics with added masses and inertias, 3D environments, disturbed operating conditions and high level of uncertainty in state knowledge and environment. Otherwise, they share qualities with typical robotic problems including partial knowledge of the environment and tasks that can be precisely specified or not. These tasks can involve continuous interaction with the environment.

The purpose of this book is to familiarize the readers with some planning and control strategies that have been proven efficient through research. It is made of a hierarchy of modules with well defined functions operating at a variety of rates, linked together from top to bottom. The outer loop, closed periodically, consists of a discrete search that produces a set of waypoints leading to the goal while avoiding obstacles and weighed regions. The second level smoothes this set so that the generated paths are feasible and the last one is the tracking controller that attempts to minimize the error between the robot measured trajectory and the reference trajectory.

This hierarchy conveys to the content of the book: Modeling, Mission Planning, Trajectory Design and Control.

Paris, France

Yasmina Bestaoui Sebbane

Contents

Abbreviations

Acronyms	
2D	Two Dimensional
3D	Three Dimensional
ADS-B	Automatic Dependent Surveillance Broadcast
DCM	Direction Cosine Matrix
DME	Distance Measurement Equipment
DMOC	Discrete Mechanics and Optimal Control
DOF	Degrees Of Freedom
EFCS	Electronic Flight Control System
ENU	East North Up frame
ESD	Event Sequence Diagram
EST	Expansive Search Trees
FDIR	Fault Detection Isolation and Response
FIFO	First In First Out
FFA	Functional Fault Analysis
FMEA	Failure Modes Effects Analysis
FMS	Flight Management System
FT	Fault Tree
FTA	Fault Tree Analysis
GCD	Greater Common Divisor
GIS	Geographic Information System
GNC	Guidance, Navigation and Control
GNSS	Global Navigation Satellite System
GPS	Global Positioning System
HTA	Heavier Than Air
ILS	Instrument Landing System
IMU	Inertial Measurement Unit
LIFO	Last In First Out
LMI	Linear Matrix Inequality
LQE	Linear Quadratic Estimator
LQR	Linear Quadratic Regulator

LQG	Linear Quadratic Gaussian
LRU	Line Replaceable Unit
LTA	Lighter Than Air
LTAR	Lighter Than Air Robot
MA	Maneuver Automaton
MDL	Motion Description Language
MDP	Markov Decision Process
MILP	Mixed Integer Linear Programming
MIMO	Multiple Input Multiple Output system
MLD	Master Logic Diagram
MMAE	Multiple Model Adaptive Estimation
MPC	Model Predictive Control
NED	North East Down frame
NP	Non Polynomial
PID	Proportional Integral Derivative
PP	Pole Placement
PRA	Probabilistic Risk Assessment
PRM	Probabilistic Road Map
RHC	Receding Horizon Control
RHTA	Receding Horizon Task Assignment
RRT	Rapidly exploring Random Tree
SHM	System Health Management
SVD	Singular Value Decomposition
UAV	Unmanned Aerial Vehicle
VFF	Virtual Force Field
VOR	VHF Omni-directional Range
VSC	Variable Structure Control

Nomenclature

α	Angle of attack
β	Side-slip angle
χ	Heading angle
η	6D vector position and orientation of the body fixed frame
$\eta_1 = (x, y, z)^T$	3D position of the body fixed frame expressed in the Earth fixed frame
η_2	Orientation of the body fixed frame expressed in the Earth fixed frame, either expressed with Euler angles $(\phi, \theta, \psi)^T$ or Euler parameters $(q_0, q_1, q_2, q_3)^T$
γ	Flight path angle
γ_1	Upper bound of the perturbed closed loop transfer matrix
Γ	Vortex intensity
λ	Lagrange multiplier
λ_g	Goal sink strength
μ	Tilt angle of the main propellers
ω_a	Solid body rotation speed of the air
$\Omega = (p, q, r)^T$	Angular velocity expressed in the LTAR fixed frame

ϕ	Roll angle
Φ	Scalar velocity potential
Φ_u	uniform flow
Π	Kinetic moment
ψ	Yaw angle
ρ	Volume mass of the surrounding fluid
θ	Pitch angle
A	State matrix
A_{lat}, B_{lat}	Lateral state and control matrices
A_{long}, B_{long}	Longitudinal state and control matrices
B	Control Matrix
b	Fin semi span
B(s)	Binormal vector
C_3	Anti-symmetric cross product matrix
C_T	Tangential aerodynamic force coefficient
C_N	Normal aerodynamic force coefficient
C_L	Lateral aerodynamic force coefficient
C_l	Tangential aerodynamic torque coefficient
C_m	Normal aerodynamic torque coefficient
C_n	Lateral aerodynamic torque coefficient
d	Distance
E_p	Potential energy
F	Force acting on the vehicle
F_m	Main torque
F_T	Tail torque
g	Acceleration of the gravity
G	Center of gravity of the airship
H	Hamiltonian function
$I_{n\times n}$	$n \times n$ identity matrix
J	Cost function
J	Inertia matrix
K	Control gain matrix
L	Gain matrix of the linear quadratic estimator
m	Mass
M	Moment acting on the vehicle
M	Generalized mass matrix
M_a	Added mass matrix
N	Bow of the airship
$\mathbf{N}(s)$	Normal vector
$O_{n\times m}$	$n \times m$ zero matrix
on	Number of obstacles
P	Pressure
P_d	Power of the dissipative forces
q	Quaternion
Q	Configuration

Q_g	Goal configuration
Q_1	State weighting matrix
R_s	Hull cross sectional radius
R_v	Ring vortex radius
R	3 × 3 Rotation matrix
$\mathcal{R}_a$	Aerodynamic frame
$\mathcal{R}_f$	Inertial fixed frame
$\mathcal{R}_m$	Body fixed frame
R_j	Distance of the current point to an arbitrary point on the spatial panel j
R	6 × 6 Generalized Rotation matrix
R_1	Control input weighting matrix
s	Curvilinear abscissa
S	Riccati matrix
S_{ref}	Characteristic or reference or wetted area
S_j	Surface of the panel j
$Sk(V)$	Skew matrix related to the vector V
$SO(3)$	Special orthogonal matrix group
T	Thrust
T_A	Fluid kinetic energy
$\mathbf{T}(s)$	Tangent vector
U	Potential energy
U	Strength of the harmonic potential
$V = (u, v, w)^T$	Linear velocity of the Vehicle expressed in the LTAR fixed frame
V	Generalized velocity of the Vehicle $\mathbf{V} = (V, \Omega)^T$
V	Vectorial velocity field
V_a	Relative Vehicle airspeed
W	3D wind speed
W	Kinetic energy
W_L	Lyapunov Function
$\mathbf{W}_N$	Spectral density matrix

List of Figures

Chapter 1
Introduction

Abstract An aerial robot is a system capable of sustained flight with no direct human control and able to perform a specific task. A Lighter Than Air Robot (LTAR) is an unmanned lighter than air vehicle with sufficient autonomy. Robotic airships can also be called Aerobot. Lighter-Than-Air-Vehicles are becoming more popular in aerial missions. A Lighter Than Air Robot behaves and functions differently to a heavier than air craft: airplane and helicopter. It does not require any motor action to maintain a certain altitude and position in the space as it relies on low density gas inside the envelope to balance its own weight. It uses buoyancy to float in the air.

1.1 Aerial Robotics

Aerial robotics is a growing field with many application possibilities. Aerial robots can be viewed as the evolution of the unmanned aerial vehicles. An unmanned System is defined as an electro-mechanical system, with no human operator aboard, that is able to exert its power to perform designed missions. An autonomous system reacts to its external inputs and takes some action without operator control. The goal is to design a system that reacts to its environment and plans its own activities to achieve the mission goals.

An unmanned system may be mobile or stationary and it includes categories of unmanned ground vehicles, unmanned underwater vehicles, unmanned surface vehicles and unmanned aerial vehicles (UAV). Unmanned Aerial Vehicles are divided into three categories: rotary wings (helicopter), fixed wing (airplane) and lighter than air (airship). An airship is an engine-driven lighter-than-air craft that can be steered. Modern airships employ advanced technologies such as composite materials, modern electronic systems, fly by light or fly by wire transmissions controls and the latest theories in stability, control and aerodynamics. Some aspects of modern design are borrowed from other disciplines, while in certain areas such as envelope fabrics and automatic docking systems, the technology is specifically developed for airships. The conventional airship is essentially a low speed vehicle with the power requirement being approximately proportional to the cube of the airspeed. An airship is very fuel efficient, and its endurance, one of its primary benefits, can be several orders of magnitude greater than for a heavier than air craft. The payload

Y. Bestaoui Sebbane, *Lighter than Air Robots*,
Intelligent Systems, Control and Automation: Science and Engineering 58,
DOI 10.1007/978-94-007-2663-5_1,

of airships is generally limited by the gas lift available in the climatic conditions prevailing at the cruise flight altitude.

A lighter than air robot is an unmanned airship with sufficient autonomy. Vehicle autonomy is a discipline fertilized by the robotics and computer science fields. Autonomy implies a degree of self-regulation inherent in a system's operation and is a key technology for future high-performance and remote operation applications. The promise of autonomy includes

- Reduced need for human intervention, which is specially important for dull and/or dangerous missions
- Increased performance range and capabilities
- Extended operation life
- Decreased cost

The requirement for autonomy from direct human interaction places heavy emphasis on reliable guidance and control strategies. Three main layers are identified in the context of autonomous aerial vehicles [71, 72]:

- Strategic layer: This layer includes a number of functions which are responsible to assess the current status of the system and the mission planning with the goal of defining and prioritizing the mission objectives. Generation of way-points and task scheduling are examples of activities in this layer: topic of Chap. 3.
- Tactical layer: This layer includes functions which aim at devising plans to achieve the goals set by the strategic layer using the available resources. The function of guidance is included in this layer: topic of Chap. 4.
- Skill layer: This layer includes all the functions performed by the control system to execute the plan devised by the tactical layer such as control and regulation: topic of Chap. 5.

For an unmanned airship to be a lighter than air robot, it must be able to interact with the world it is working in. One such interaction is the movement which allows the lighter than air robot to move from one initial configuration to a final configuration all by itself, without the help of human and without the collision with any obstacle. The planning and control must be studied in order to make this interaction possible. The former is to find a feasible and even optimal trajectory for the robot to follow while the latter is to implement the plan [162].

Like all flying vehicles, the performance of aerial robots depends extensively on their size and the characteristics of their lifting mechanisms (wings, rotors). Due to the special requirement of automatic flight in complex outdoor environment, guidance and control are very important in practice. The specific characteristics of various vehicle types pose different challenges to the planning and control. For example, a helicopter or an airship (with a tail rotor) have the ability to stop and go backwards whereas a fixed wing aircraft has to maintain a minimum velocity, in order to fly [40, 61, 62, 79, 100, 127].

One way to deflect some of the concerns associated with high fuel consumption of heavier than air craft is to rely on lighter than air vehicles. They offer an unmatched capability to fly for long periods of time and to do it silently. Airships float in the atmosphere due to their buoyancy so are airborne when filled with gas, before being launched and remain airborne until the ballonets are filled with air or the gas eventually is released in extreme situations. With this feature, a lighter than air robot could conduct a continuous aerial operation with very low energy consumption.

Lighter than air vehicles are an attractive solution for many applications requiring a sustained airborne presence. The missions foreseen are environmental monitoring, inspection, surveillance and other applications, such that the lighter than air robot should fly at low altitude, be able to hover above points of interest, take-off and land vertically, present a secure degradation in case of failure [5, 60, 102, 108, 116]. Another application is a new concept consisting of using autonomous airships as platforms operating for extended periods of time at very high altitudes (between 20,000 and 100,000 m) to fulfill missions currently accomplished using spacecraft.

Wide area surveillance for months at a time is presently impossible as neither satellites nor aircraft can provide these capabilities simultaneously. However, renewable energy technology has progressed enough to seriously consider building an airship for ultra-long duration flights (such as thin film photo-voltaic arrays and fuel cells...) [26, 163, 175, 205, 227].

Interest in near-space solutions for telecommunications has also increased because they are more feasible than in the past, due to technological advances in lightweight materials and in solar power technology. As airships need less energy to stay aloft, they are considered as potential near-space platforms [126, 152, 153, 182].

Unmanned airships have the following modes of control: fully autonomous, semi-autonomous, tele-operation and remote control.

Fully autonomous operation is the subject of this book. Guidance, control and navigation of the lighter than air robot is accomplished by an on board self-contained flight-management control system. The aerial robot must be equipped with reliable position and actuation equipment so as to be able of controlled flight and this constitutes a requirement prior to doing research or development in this field [116, 120, 129, 138, 161, 176, 196]. The choice of sensors is critical to obtaining a properly aerial robot. Usually the same suite of sensors may not apply to all phases of flight. The individual inputs collected from each sensor are usually not sufficient to estimate the state of the vehicle. Different sensors may be efficient over different flight regimes. The proper way to leverage individual information provided by each sensor is through an appropriate filtering process that can yield rather comprehensive information about the entire system's state. The necessity for good state estimates arises early in the development process since closed loop flight would be impossible otherwise.

Lighter than air robots have essentially two flight regimes: hover and cruising flight:

- *Aerostatics hovering (Hover)*: The speed of the lighter than air robot relative to the surrounding air is small such that few or no forces act on the vehicle, except those resulting from the propulsion itself. The ability to hover is extremely useful for any operations that require a constant position versus the Earth.

- *Aerodynamics flight (Cruising Flight)*: There is a significant relative speed between the vehicle and its surrounding environment. Significant aerodynamics forces act on the vehicle. These aerodynamic forces then largely dominate those generated by the power system. Unlike hovering flight, cruising flight usually results in the aerial robot constantly meeting wind, which makes the range of adverse events to flight quite narrower.

Guidance is the logic that issues steering commands to the vehicle to accomplish certain flight objectives. The guidance algorithm generates the autopilot commands that steer the lighter than air robot. Planning trajectories is a fundamental aspect of autonomous vehicle guidance. In the literature [34, 69, 96, 101, 107, 114, 124, 132, 155, 160, 174, 199], a large number of techniques have been proposed to deal with mobile robots governed by kinematic equations and subject to non holonomic constraints. The majority of these techniques exploits the inherent structure of those systems and leads to very restricted algorithms when applied to aerial robots.

Guidance and control are the main topics of this book. It is necessary, for this purpose, to use absolute navigation and positioning that allows to obtain global and time-independent information. Inertial measurements are essential for stability control, but navigation and other high level tasks demand an additional global localization system. A typical solution adopted for aerial vehicles is the use of GPS, but the precision and the availability can appear too poor in particular flight configurations, as well as the time of delivery and processing too long. Other systems are then necessary to insure an accurate navigation at full time, coupled to the localization device. To navigate accurately, series of predefined way-points can be delivered, either by aeronautic active radio devices or by extraction of particular objects of the environment or by the mission planner. In the case of a lighter than air robot, this operation must handle meteorological information, especially force and direction of the wind, in order to plan trajectory corrections but also for a secure auto-piloting. Many sensing options exist [202, 211, 213]:

- Inertial navigation system: It consists of a combination of three orthogonally mounted accelerometers and three orthogonally mounted gyroscopes. The accelerometers measure, up to sensor error, the accelerations experienced by the vehicle at the location of the inertial sensor minus gravity. Gyroscopes measure vehicle angular velocities. Current inertial sensors are usually rigidly linked to the vehicle in order to form a strap down inertial measurement system. Inertial measurement units are extremely useful for measuring variations in acceleration and angular velocities. They constitute a basis of inner-loop control system. Practical inertial navigation system on a lighter than air robot typically receive at least position updates from other sensors called inertial aiding.
- Global Navigation system: GPS, GLONASS, GALILEO space based system offer real-time absolute position information using a constellation of satellites circumnavigating the Earth. Wherever they are available, satellite navigation systems are a convenient and cheap means for a vehicle to locate itself approximately.
- Air data probes and altimeters: Pressure measuring devices are useful sensors in airships. It is possible to measure the atmospheric pressure at the location of the

lighter than air robot and the dynamic pressure along all vehicle axes, with a suitable arrangement of pressure sensors (such as Pitot tubes). These data can themselves be transformed into useful informations about the lighter than air robot altitude and direction of motion relative to the air. Indeed, pressure probes are very sensitive to flow perturbations generated by the hull, gondola, engines and propellers. Air data probes are therefore positioned as far away as possible from the main elements of the vehicle. Mounted together with inertial measurement system, air data probes allow the airship to maintain stable flight at a prescribed altitude.
- Terrestrial radio navigation system: VHF omni-directional range (VOR)
- Weather Radar and passive vision sensors
- Magnetic compasses
- Distance measuring equipment: altitude radars, ultrasonic sensors, laser range finders...

The assumption is made throughout this book that the information required on the vehicle and the environment is available.

1.2 Outline of the Book

The contents of the book are organized as follows:

- To guide and control a lighter than air robot, useful models of their kinematics and dynamics are needed in different forms, depending on the desired type of guidance and control. The purpose of the second chapter is to present the complex flight dynamics associated with lighter than air vehicles. Kinematics are first presented using both approaches of Euler angles and Euler parameters. Then six degrees of freedom dynamics are proposed respectively within Newton-Euler and Lagrange-Hamilton frameworks followed by translational dynamics. As far as planning and guidance are concerned, translational dynamics are sufficient to represent the behavior of a lighter than air vehicle. Finally, some aerology characteristics are mentioned.
- A basic problem which has to be solved by aerial robots is the generation of a plan for moving from one location to another location in space to accomplish a desired task while at the same time avoiding collisions with obstacles. The purpose of the third chapter is to present mission planning with its subdivisions: flight planning, obstacle and collision avoidance. An overview of existing mission planning algorithms, without and with differential constraints, is provided in this chapter. The world, configuration, or state space is divided into free space and obstacle space: these are disjoint subsets of the space. Free space is the subset of points in which the vehicle can move without contact with the obstacles and obstacle space is the subset of points representing a collision between the vehicle and an obstacle. Planning with uncertain winds followed by planning with strong winds are then presented. Finally, task assignment is considered.

- The purpose of the fourth chapter is trajectory design in hover and in cruising flight. The role of the trajectory generator is to generate a feasible time trajectory for the aerial vehicle. The first section of this chapter presents trim trajectories in hover. The second section considers lateral planning of the lighter than air robot while the third one considers 3D optimal time trajectories, in the presence of bounds on the acceleration inputs considering the nominal deterministic part of the wind. The problem is in fact the determination of a control law that transfers this dynamic system between two assigned states so as to minimize the execution time. A powerful tool for solving this problem is the Pontryagin minimum principle that provides necessary conditions for optimality. By exploiting these conditions in conjunction with the analysis of the specific characteristics of the kinematics of the lighter than air robot, it is possible to identify a reduced set of candidate trajectories among which there is certainly the desired optimal solution (if it exists). The subject of the last section of this chapter is parametric curves, the lighter than air robot being subject to nonholonomic constraints. When minimizing curvature (and torsion), the problem can be solved by planning a path for the kinematic model. If the cost criterion depends on the path as well as the timing law, it is necessary to plan directly on the kinematic model. The dynamic model is used for the computation of the various constraints on velocities and higher derivatives.
- The purpose of the fifth chapter is the design of advanced control system, using respectively the linear approach in the second section and the nonlinear approaches in the third section. The control characteristics of the vehicle have to be evaluated by considering specific tasks such as ability to maneuver from hover, to accelerate into a heavy wind or cross wind, and to hover a point on the ground in a variable, shifting wind... Subsequently, the control power characteristics of the vehicle are to be determined by considering the proposed control concepts. Classical linear control and linear robust control are briefly presented in the first part of the chapter. In the second part of the chapter, some nonlinear approaches such as dynamic inversion, sliding mode and backstepping controllers are then considered. To present a more complete presentation, diagnosis and response to systems failure must also be considered. The selection of the performance model incorporating health state information about the vehicle is also an important aspect of the feedback design.
- Finally, general conclusions and perspectives are presented in the last chapter and some current projects are proposed in the Appendix.

Chapter 2
Modeling

Abstract Nowadays, non rigid airships with a cigar shaped profile are the most common type. These airships do not have any rigid internal framework. The objective of this chapter is to present kinematics and dynamics models of a lighter than air robot, taking into account wind effect. Newton-Euler and Hamilton-Lagrange approaches are used for this discussion then the translational model is presented. Here, motion is referenced to a system of orthogonal body axes fixed in the airship, with the origin assumed to coincide with the bow. Finally, some aerology characteristics are briefly discussed.

2.1 Introduction

The airship is essentially a low speed vehicle and is fundamentally a balloon with a means of propulsion but in order to produce an effective vehicle, the balloon element has to be considerably modified. The primary requirement of the envelope is that it should be of streamline form to reduce air resistance and improve controllability and that it should be able to sustain the additional loads imposed by aerodynamic forces and propulsive components. A representative mathematical model of a lighter than air vehicle differs from the usual aircraft model because the lighter than air vehicle displaces a very large volume and its virtual mass and inertia properties become significant and thus the aerodynamics and control characteristics are described quite differently.

What makes a vehicle lighter than air is the fact that it uses a lifting gas (i.e. Helium as Hydrogen has been banned) in order to be lighter than the surrounding air. While Hydrogen gas is approximately 7 percent more buoyant, Helium has the advantage of being non-inflammable in addition to being fire retardant.

The principle of Archimedes applies in the air as well as under water. The difference between airships and balloons is that: balloons simply follow the direction of the winds. In contrast, airships are powered and have some means of controlling their direction. There are basically three different kinds of airships: rigid, semi-rigid and non-rigid airships (see Fig. 2.1).

Non rigid airships or pressure airships, also known as blimps, are the most common form nowadays. They are basically large gas balloons. Their shape is maintained by their internal overpressure. The only solid parts are the gondola, the set of

Y. Bestaoui Sebbane, *Lighter than Air Robots*,
Intelligent Systems, Control and Automation: Science and Engineering 58,
DOI 10.1007/978-94-007-2663-5_2, © Springer Science+Business Media B.V. 2012

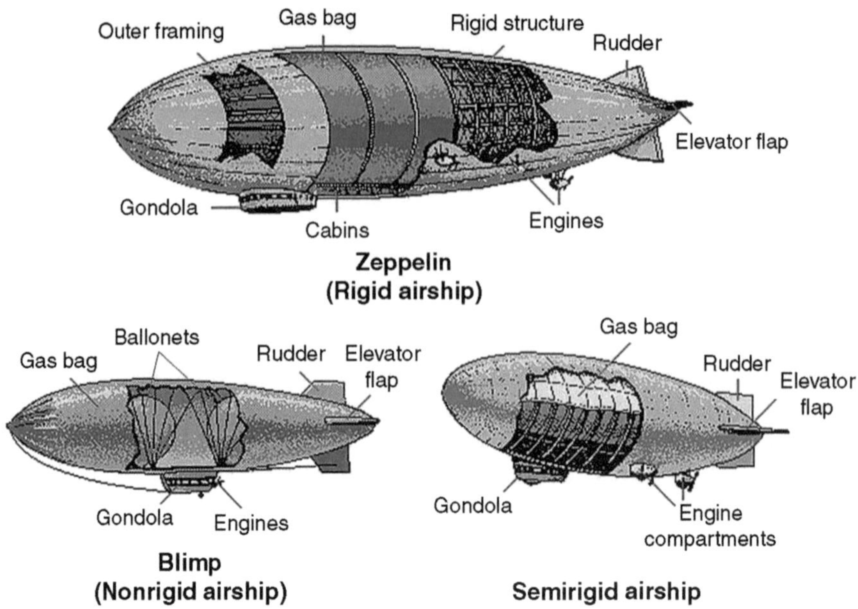

Fig. 2.1 Different kind of *airships*, taken from http://library.thinkquest.org/18033/airship.html

propellers and the tail fins. The envelope holds the Helium that makes the airship lighter than air. In addition to the lift provided by Helium, airships derive aerodynamic lift from the shape of the envelope as it moves through the air. An envelope as the gas containment membrane encloses the lifting gas and the ballonets and provides protection from the environment. Ballonets are filled with air by blowers to maintain a fixed pressure inside as the temperature of the lifting gas or the airship altitude changes. Ballonets permit the envelope pressure to be controlled and relative fullness of fore and aft ballonets is associated with pitch control. Adjustment of air volume in ballonets and gas volume in the envelope produces the change of buoyancy. The vertical portion of the gondola loads is supported by an internal suspension system which is contained in the envelope and runs from the top of the envelope to the gondola. The principal function of the external suspension system attached to the bottom part of the gondola loads into the envelope. The envelope fabric consists of laminated composite and is designed to withstand environment and flight loads. Non rigid airships have simple structures and are easy to design, build and maintain.

The most common form of a dirigible is like an ellipsoid. It is a highly aerodynamically profile with a good resistance to aerostatics pressures. Aerodynamics forces occur when a body moves relative to the air in which it is immersed. These forces are classified as steady when they are invariant with time or transient when they are not. The steady forces arise when a body is in uniform motion in a still atmosphere. Transient forces occur during maneuvers or owing to turbulence in the atmosphere. Aerodynamic forces and moments arise from the local surface pressures, which when integrated over the whole body, give the overall forces and moments. It is therefore necessary to determine the local pressures. At non-neutral lift, it becomes necessary to trim the airship to generate aerodynamic lift to balance the difference between the weight and lift forces. This in turn creates additional drag with the consequent penalty in performance.

Several research studies have been carried out on dynamic modeling and simulation of airships. Dynamic models are built to study guidance and control techniques of airships, trajectory, stability and maneuverability. This chapter consists of 5 sections. Section 2.2 presents kinematics using two different approaches: Euler angles and Euler parameters. In Sect. 2.3, six degrees of freedom dynamics are presented, using respectively Newton-Euler and Lagrange-Hamilton methods, followed by three degrees of freedom translational dynamics and aerology characteristics, in Sect. 2.4. Finally some concluding remarks are given in Sect. 2.5.

2.2 Kinematics

The model used was written originally for a buoyant underwater vehicle [73]. It was modified later to take into account the specificity of the airship [10, 15, 23, 24, 89, 108]. The position of all points belonging to the airship rigid body with respect to the inertial fixed frame $\mathcal{R}_f$ can be completely defined by knowing the orientation of a body fixed frame $\mathcal{R}_m$ to the airship body and the position of its origin with respect to the reference frame $\mathcal{R}_f$. The choice of the body fixed frame origin can be made at the center of gravity, center of volume (assumed to coincide with the gross center of lift) or also at the bow of the hull. The Aerodynamic frame $\mathcal{R}_a$ is used extensively in flight mechanics with flight equations of motion and through an aircraft's air data probe and other sensors. The airship's air speed, V_a is referenced to the aerodynamic frame. These three frames are classically considered in the derivation of the kinematical and dynamical equations of the motion of the aerial vehicles. There are many ways to describe finite rotations between frames. Direction cosines, quaternions, Euler angles, can serve as examples. Some of these groups of variables are very close to each other in their nature.

Remark 2.1 In classical vector mechanics, it is customary to represent the acceleration of a rigid body by means of the linear acceleration of a specified point in the body and an angular acceleration vector which applies to the whole body.

Remark 2.2 The derivation relationship in two different frames can be written as

$$V = \dot{r} + \Omega \times r \tag{2.1}$$

$$\dot{V} = \ddot{r} + \dot{\Omega} \times r + \Omega \times (\Omega \times r) \tag{2.2}$$

where Ω is the angular velocity.

Remark 2.3 The finite rotation of a rigid body does not obey to the laws of vector addition (in particular commutativity) and as a result the angular velocity of the body cannot be integrated to give the attitude of a body [13].

These remarks are extensively used in the following demonstrations.

2.2.1 Euler Angles

The usual minimal representation of orientation is given by a set of three Euler angles. Assembled with the three position coordinates, they allow the description of the situation of a rigid body. A 3×3 direction cosine matrix (of Euler parameters) is used to describe the orientation of the body (achieved by 3 successive rotations) with respect to some fixed frame reference. This parametrization is a three parameters set corresponding to the yaw ψ, pitch θ and roll ϕ, which can be directly measured by the attitude sensors. Adopting this formulation, the rotation matrix R can be written as a function of $\eta_2 = (\phi, \theta, \psi)^T$ given by:

$$R(\eta_2) = R_z(\psi)R_y(\theta)R_x(\phi) \tag{2.3}$$

with

$$R_x(\phi) = \begin{pmatrix} 1 & 0 & 0 \\ 0 & \cos\phi & -\sin\phi \\ 0 & \sin\phi & \cos\phi \end{pmatrix}$$

$$R_y(\theta) = \begin{pmatrix} \cos\theta & 0 & \sin\theta \\ 0 & 1 & 0 \\ -\sin\theta & 0 & \cos\theta \end{pmatrix}$$

$$R_z(\psi) = \begin{pmatrix} \cos\psi & -\sin\psi & 0 \\ \sin\psi & \cos\psi & 0 \\ 0 & 0 & 1 \end{pmatrix}$$

This transformation, also called the Direction Cosine Matrix (DCM), can be explicited as:

$$R(\eta_2) = \begin{pmatrix} \cos\psi\cos\theta & -\sin\psi\cos\phi + \cos\psi\sin\theta\sin\phi & \sin\psi\sin\phi + \cos\psi\sin\theta\cos\phi \\ \sin\psi\cos\theta & \cos\psi\cos\phi + \sin\psi\sin\theta\sin\phi & -\cos\psi\sin\phi + \sin\psi\sin\theta\cos\phi \\ -\sin\theta & \cos\theta\sin\phi & \cos\theta\cos\phi \end{pmatrix} \tag{2.4}$$

The rotation matrix R is an orthogonal matrix. The set of such matrices constitutes the special orthogonal matrix group $SO(3)$ defined as:

$$SO(3) = \left\{ R \in \Re^{3\times 3}, R^T R = I_{3\times 3}, \det(R) = 1 \right\} \tag{2.5}$$

The relationship between the body fixed angular velocity vector $\nu_2 = (p, q, r)^T$ and the rate of change of the Euler angles $\dot{\eta}_2 = (\dot{\phi}, \dot{\theta}, \dot{\psi})^T$ can be determined by resolving the Euler rates into the body fixed coordinate frame:

$$\begin{pmatrix} p \\ q \\ r \end{pmatrix} = \begin{pmatrix} \dot{\phi} \\ 0 \\ 0 \end{pmatrix} + \begin{pmatrix} 1 & 0 & 0 \\ 0 & \cos\phi & \sin\phi \\ 0 & -\sin\phi & \cos\phi \end{pmatrix} \begin{pmatrix} 0 \\ \dot{\theta} \\ 0 \end{pmatrix} + \begin{pmatrix} 1 & 0 & 0 \\ 0 & \cos\phi & \sin\phi \\ 0 & -\sin\phi & \cos\phi \end{pmatrix} \begin{pmatrix} \cos\theta & 0 & -\sin\theta \\ 0 & 1 & 0 \\ \sin\theta & 0 & \cos\theta \end{pmatrix} \begin{pmatrix} 0 \\ 0 \\ \dot{\psi} \end{pmatrix} \tag{2.6}$$

or equivalently

$$\begin{aligned} p &= \dot{\phi} - \dot{\psi} \sin\theta \\ q &= \dot{\theta} \cos\phi + \dot{\psi} \sin\phi \cos\theta \\ r &= -\dot{\theta} \sin\phi + \dot{\psi} \cos\phi \cos\theta \end{aligned} \tag{2.7}$$

The configuration of the airship is completely defined by associating the orientation matrix and the $\mathcal{R}_m$ frame origin position vector, $\eta_1 = (x, y, z)^T$, with respect to $\mathcal{R}_f$ using homogeneous matrix formulation as:

$$A_M = \begin{pmatrix} R(\eta_2) & \eta_1 \\ 0_{3\times 3} & 1 \end{pmatrix} \tag{2.8}$$

The set of all such matrices is called the special Euclidean group of rigid-body transformations in three dimensions noted $SE(3)$ defined by:

$$SE(3) = \left\{ A_M \middle| A_M = \begin{pmatrix} R(\eta_2) & \eta_1 \\ 0_{3\times 3} & 1 \end{pmatrix}, R \in SO(3), \eta_1 \in \Re^3 \right\} \tag{2.9}$$

The set $SE(3)$ is a Lie group. Given a curve $C(t) : [-a, a] \to SE(3)$, an element $S(t)$ of the Lie algebra $\mathrm{se}(3)$ can be associated to the tangent vector $\dot{C}(t)$ at an arbitrary configuration $A(t)$ by:

$$S(t) = A_M^{-1}(t)\dot{A}_M(t) = \begin{pmatrix} Sk(\nu_2) & R^T\dot{\eta}_1 \\ 0 & 0 \end{pmatrix} \tag{2.10}$$

where

$$Sk(\nu_2) = R^T(t)\dot{R}(t) \tag{2.11}$$

is a 3×3 skew symmetric operator on a vector defined by:

$$Sk(\nu_2) = Sk((p, q, r)^T) = \begin{pmatrix} 0 & -r & q \\ r & 0 & -p \\ -q & p & 0 \end{pmatrix} \tag{2.12}$$

such that

$$\forall x, y \in \Re^3: \quad Sk(y)x = y \times x \tag{2.13}$$

A curve on $SE(3)$ physically represents a motion of the rigid body. If $(\omega(t), \nu(t))$ is the pair corresponding to $S(t)$, then $\nu_2 = \Omega = (p, q, r)^T$ physically corresponds to the angular velocity of the rigid body, while $\nu_1 = V = (u, v, w)^T$ is the linear velocity of the origin O_m of the frame $\mathcal{R}_m$ with respect to the inertial frame $\mathcal{R}_f$, both the linear and angular velocities are expressed in the body fixed frame $\mathcal{R}_m$.

The kinematic relationship between the different velocities are given by:

$$\begin{pmatrix} \dot{\eta}_1 \\ \dot{\eta}_2 \end{pmatrix} = \mathbf{RV} = \begin{pmatrix} R(\eta_2) & 0_{3\times 3} \\ 0_{3\times 3} & J(\eta_2) \end{pmatrix} \begin{pmatrix} V \\ \Omega \end{pmatrix} \tag{2.14}$$

where R is the rotation matrix defined by Eq. (2.4) and $\mathbf{J}(\boldsymbol{\eta}_2)$ is defined by:

$$J(\eta_2) = \begin{pmatrix} 1 & 0 & -\sin\theta \\ 0 & \cos\phi & \sin\phi\cos\theta \\ 0 & -\sin\phi & \cos\phi\cos\theta \end{pmatrix}^{-1} = \begin{pmatrix} 1 & \sin\phi\tan\theta & \cos\phi\tan\theta \\ 0 & \cos\phi & -\sin\phi \\ 0 & \frac{\sin\phi}{\cos\theta} & \frac{\cos\phi}{\cos\theta} \end{pmatrix} \tag{2.15}$$

This matrix presents a singularity for $\theta = \pm\frac{\pi}{2}$. During practical operations of the airship, the pitch angle of $\theta = \pm\frac{\pi}{2}$ is not likely to be encountered.

2.2.2 *Euler Parameters*

To avoid the singularity inherent to this representation, Euler parameters can be presented. They are unit quaternions and are represented by a normalized vector of four real numbers [7, 9, 15, 24]. Taking the vector part of a unit quaternion and normalizing it gives the rotational axis and the scalar part allows to obtain the angle of rotation.

2.2.2.1 Introduction

A general spatial displacement of a rigid body consists of a finite rotation about a spatial axis and a finite translation along some vector. The rotational and translational axes in general are not related to each other. It is often easier to describe a spatial displacement as a combination of a rotation and a translation motions, where the two axes are not related. However, the combined effect of the two partial transformations can be expressed as an equivalent unique screw displacement. The concept of a screw thus represents an ideal mathematical tool to analyze spatial transformation [177].

The Euler representation, a four parameter representation, is based on the idea that a transformation from one co-ordinate frame to another may be effected by a single rotation about a vector μ defined with respect to the reference frame. Let $\mathbf{q}$ denote the Euler parameters which are expressed by the rotation axis n and the rotation angle μ about the axis as follows:

$$\mathbf{q} = \begin{pmatrix} q_0 \\ q_1 \\ q_2 \\ q_3 \end{pmatrix} = \begin{pmatrix} \cos(\frac{\mu}{2}) \\ \sin(\frac{\mu}{2})n_x \\ \sin(\frac{\mu}{2})n_y \\ \sin(\frac{\mu}{2})n_z \end{pmatrix}, \quad 0 \le \mu \le 2\pi \tag{2.16}$$

The quaternion can also be written as:

$$\mathbf{q} = q_0 + q_1\hat{i} + q_2\hat{j} + q_3\hat{k} \tag{2.17}$$

q_0 is real, q_1, q_2, q_3 are imaginary. The following properties apply:

$$\begin{array}{ccc} \hat{i}.\hat{i} = -1 & \hat{j}.\hat{j} = -1 & \hat{k}.\hat{k} = -1 \\ \hat{i}.\hat{j} = -\hat{j}.\hat{i} = \hat{k} & \hat{k}.\hat{i} = -\hat{i}.\hat{k} = \hat{j} & \hat{k}.\hat{j} = -\hat{j}.\hat{k} = -\hat{i} \end{array} \tag{2.18}$$

The quaternion product may be expressed in matrix form as:

$$\mathbf{q}.\mathbf{p} = \begin{pmatrix} q_0 & -q_1 & -q_2 & -q_3 \\ q_1 & q_0 & -q_3 & q_2 \\ q_2 & q_3 & q_0 & -q_1 \\ q_3 & -q_2 & q_1 & q_0 \end{pmatrix} \begin{pmatrix} p_0 \\ p_1 \\ p_2 \\ p_3 \end{pmatrix} \tag{2.19}$$

2.2.2.2 Propagation of Quaternion with Time

The configuration space $R^3 \times SO_3$ may be replaced by $R^3 \times S^3$ where

$$S^3 = \left\{ x \in R^4 \text{ such that } \|x\|_2 = 1 \right\}$$

is the unit sphere in R^4 for the Euclidean norm $\| \cdot \|_2$. The associated kinematics differential equations of Euler parameters define a vector field on S^3 (the 3 sphere on R^4) and are given by:

$$\begin{aligned} \dot{q} &= \frac{1}{2}\left(q_0 I_{3\times 3} + Sk(q)\right)\Omega = \frac{1}{2}\left(q_0\Omega - \Omega \times q\right) \\ \dot{q}_0 &= -\frac{1}{2} q^T \Omega \end{aligned} \tag{2.20}$$

where q is the vectorial part of the quaternion. Thus, the time derivative of Euler parameters can be determined at any configuration of the body if the angular velocity is given.

Remark 2.4 Equation of this form may be solved in a navigation system to keep track of the quaternion parameters which define body orientation. The quaternion parameters may then be used to compute an equivalent direction cosine matrix, or used directly to transform the measured specific force vector into the chosen reference frame.

2.2.2.3 Kinematic Model

Let us introduce $\eta = (\eta_1, \eta_2)^T$ where $\eta_1 = (x, y, z)^T$ is the position vector of the airship (expressed in the Earth fixed frame), $\eta_2 = (q_0, q_1, q_2, q_3)^T$ is the quaternion representing the orientation, V as the linear velocity of the origin and Ω as the angular velocity (expressed in the airship fixed frame):

$$V = \begin{pmatrix} u \\ v \\ w \end{pmatrix} \qquad \Omega = \begin{pmatrix} p \\ q \\ r \end{pmatrix} \tag{2.21}$$

The kinematics of the lighter than air robot can be expressed in the following way:

$$\dot{\eta} = \begin{pmatrix} \dot{\eta}_1 \\ \dot{\eta}_2 \end{pmatrix} = \begin{pmatrix} R(\eta_2) & 0 \\ 0 & J(\eta_2) \end{pmatrix} \mathbf{V} = \mathbf{R}\mathbf{V} \tag{2.22}$$

where $\mathbf{V} = (V, \Omega)^T$.

The orientation matrices R and J are formulated as follows [167, 218]:

$$R(\eta_2) = \begin{pmatrix} 1-2(q_2^2+q_3^2) & 2(q_1q_2+q_3q_0) & 2(q_1q_3-q_2q_0) \\ 2(q_1q_2-q_3q_0) & 1-2(q_1^2+q_3^2) & 2(q_2q_3+q_1q_0) \\ 2(q_1q_3+q_2q_0) & 2(q_2q_3-q_1q_0) & 1-2(q_1^2+q_2^2) \end{pmatrix} \tag{2.23}$$

or equivalently

$$R = \left(q_0^2 - q^T q\right) I_{3\times3} + 2qq^T + 2q_0 Sk(q) \tag{2.24}$$

and

$$J(\eta_2) = \frac{1}{2} \begin{pmatrix} -q_1 & -q_2 & -q_3 \\ q_0 & -q_3 & q_2 \\ q_3 & q_0 & -q_1 \\ -q_2 & q_1 & q_0 \end{pmatrix} \tag{2.25}$$

with $J(\eta_2)^T J(\eta_2) = \frac{1}{4} I_{3\times3}$. If the following diagonal matrix is defined:

$$\boldsymbol{\Delta}_{6\times6} = \begin{pmatrix} \mathbf{I}_{3\times3} & \mathbf{0}_{3\times3} \\ \mathbf{0}_{3\times3} & \frac{1}{4}\mathbf{I}_{3\times3} \end{pmatrix} \tag{2.26}$$

Then:

$$\mathbf{R}^T \mathbf{R} = \boldsymbol{\Delta}_{6\times6} \tag{2.27}$$

giving way to the formulation for the left pseudo-inverse of $\mathbf{R}$ as

$$\mathbf{R}^{+} = \boldsymbol{\Delta}_{6\times6}^{-1} \mathbf{R}^T \tag{2.28}$$

This formulation is only achievable due to the fact that the quaternions are used instead of the Euler angles for attitude representation.

Remark 2.5 Euler parameters cannot be used in a Lagrangian approach of dynamic modeling since this 7 parameters configuration is not a minimal one.

2.3 Dynamics

A mathematical description of the flight dynamics must include the aerodynamic, structural and other internal dynamics effects that influence the response of the lighter than air robot to inputs control and external atmospheric disturbances [17, 23, 82, 89]. In this section, analytic expressions for the forces and moments on the lighter than air robot are derived. The forces and moments are referred to a system of body-fixed axes, centered at the lighter than air bow. There are in general two approaches in deriving equations of motion. One is to apply Newton-Euler's law and the other one is the Lagrange-Hamilton framework.

Remark 2.6 For preliminary analysis, an airship can be treated as a loaded beam subjected to bending forces from unequal distribution of weight and buoyancy and shear forces are calculated by dividing the airship body into many longitudinal segments. The computation of the static bending moments can be separated into three parts: vertical forces of distributed weight and buoyancy, longitudinal components of suspension rope tension and longitudinal forces due to increase of gas superpressure [133]. The contribution of the gondola weight is included in the first two parts. The aerodynamic bending moment can be computed from the effect of the empennage lift and transverse forces on the envelope (using Munk's momentum theory [154]).

The lighter than air platform considered in this book is a generic airship. It is a non rigid airship equipped with two vectorable engines on the sides of the gondola, one tail rotor and 4 control surfaces at the stern. This airship is controlled with elevator and rudder surfaces for controllability of the pitch and heading angles, respectively, at significant airspeeds. These surfaces are mounted at the rear of the airship. Bow and/or stern thrusters are also used for landing and docking operations.

Remark 2.7 Some simplifying assumptions are made in this derivation: the Earth fixed reference frame is inertial, the gravitational field is constant. The hull is considered as a solid. Elasticity is not considered in this derivation [10]. The motion of the Helium inside the hull is neglected. The volume of the lighter than air robot is considered constant. As the speed of a lighter than air robot is generally low, the coupling between dynamics and thermal phenomena are neglected and the density of the air is not locally modified by the system's motion.

2.3.1 Mass Characteristics

A significant difference of an airship from a typical aircraft is that its mass characteristics strongly depends on the change of altitude [105]. The pressure difference between the surrounding atmosphere and the inner gas (Helium) should be kept as constant as possible at each altitude level. This permanent pressure difference is required for maintaining the aerodynamic shape of the envelope under most operational conditions. As the atmospheric pressure changes with the altitude, it should be compensated by the internal pressure. For this purpose, the envelope is equipped with two air filled ballonets. The volume occupied by the inner gas and the ballonets represents the inner volume of the hull envelope and is nearly constant. By filling the ballonets with the air, they displace the volume of the inner gas, increasing the total pressure of the gas in the envelope. In general, the total mass of the airship can be expressed by the following equation:

$$m = m_{gas} + m_{Bal} + m_{ic} + m_F + m_P \tag{2.29}$$

where m_{gas} is the mass of the inner gas (Helium), m_{Bal} is the total mass of air ballonets, m_{ic} represents the mass of all internal components (skin, structures...), $m_F(t)$ the time varying fuel mass and finally m_p the payload mass. The mass of the Helium can be considered as constant if leakage through the hull's skin is insignificant. The mass of all internal components can be derived by accounting all elements of the airship as a consolidation of point and distributed masses [215, 217]. Each ballonet can be modeled as a fixed point with variable mass [42]. The volume of ballonets depends on change of the atmospheric pressure gradient. At sea level, where the atmospheric level is high, the ballonet volume has its maximum level and reduces with the increased altitude.

2.3.2 6 DOF Dynamics: Newton-Euler Approach

In this section, the dynamics model is defined as the set of equations relying the situation of the vehicle in its position, velocity and acceleration to the control vector. The forces and moments are referred to a system of body-fixed axes, centered at the bow N of the lighter than air robot.

Let the motion of the lighter than air robot be described by its velocity $\mathbf{V}$ a 6D vector including the linear V and angular Ω velocities. Let the surrounding air be described by an wind velocity V_a. The wind coordinates vector is defined as: $\eta_w = (x_w\ y_w\ z_w\ \phi_w\ \theta_w\ \psi_w)^T$ whose time derivative is relative to the wind velocity

$$\dot{\eta}_w = \mathbf{R}\mathbf{V}_w \tag{2.30}$$

The robot has thus a relative air velocity $V_a = V - V_w$. As the vector from point N, being the bow to point G, being the center of gravity, is represented by $\overline{OG} -$

$\overline{ON} + \overline{NG}$ where

$$\overline{NG} = \begin{pmatrix} x_g & 0 & z_g \end{pmatrix}^T$$

Both velocities are related via:

$$V_G = V - Sk\left(\overline{NG}\right)\Omega \tag{2.31}$$

where $Sk(\overline{NG})$ is the skew matrix related to $\overline{NG}$.

2.3.2.1 Newton-Euler Equations

Newton's laws of motion are used to relate the applied forces and moments to the resulting translational and rotational accelerations. Translational and rotational motions are considered in the sequel.

Translational Motion Concerning forces, the Newton-Euler equations of motion may be given by the following relations:

$$\sum F_{ext} = \left.\frac{dP_d}{dt}\right|_{\mathcal{R}_f} = \left.\frac{dP_d}{dt}\right|_{\mathcal{R}_m} + \Omega \times P_d \tag{2.32}$$

The generalized force vector is given by F_{ext} while the linear momentum is given by:

$$P_d = mV_G = m\left(V - Sk\left(\overline{NG}\right)\Omega\right) \tag{2.33}$$

Rotational Motion Concerning moments, the Newton-Euler equations of motion may be given by the following relations:

$$\sum M_{ext} = \left.\frac{d\Pi_N}{dt}\right|_{\mathcal{R}_f} + V \times P_d = \left.\frac{d\Pi_N}{dt}\right|_{\mathcal{R}_m} + \Omega \times \Pi_N + V \times P_d \tag{2.34}$$

with

$$\Pi_N = \Pi_G + \overline{NG} \times mV_G = I_G\Omega + \overline{NG} \times mV_G = I_N\Omega + \overline{NG} \times mV \tag{2.35}$$

where Π_N and Π_G are the angular momentum computed respectively at the points N and G. From Koenig theorem, the inertia operator in N is calculated as:

$$I_N = I_G + m\overline{NG} \times \left(\Omega \times \left(\overline{NG}\right)\right) \tag{2.36}$$

Thus, these forces and torques equations may be written as:

$$\begin{aligned} \sum F_{ext} = {} & m\dot{V}_{/\mathcal{R}_m} - mSk\left(\overline{NG}\right)\dot{\Omega}_{/\mathcal{R}_m} + \dot{m}V_{/\mathcal{R}_m} - \dot{m}Sk\left(\overline{NG}\right)\Omega_{/\mathcal{R}_m} \\ & + m\Omega \times V_{/\mathcal{R}_m} - m\Omega \times Sk\left(\overline{NG}\right)\Omega_{/\mathcal{R}_m} \end{aligned} \tag{2.37}$$

and

$$\sum M_{ext} = I_N \dot{\Omega}_{/\mathcal{R}_m} + mSk\left(\overline{NG}\right)\dot{V}_{/\mathcal{R}_m} + \Omega \times I_N \Omega \\ + \Omega \times mSk\left(\overline{NG}\right)V + M_{var} \tag{2.38}$$

where

$$M_{var} = \dot{I}_N \Omega_{/\mathcal{R}_m} + \dot{m}Sk\left(\overline{NG}\right)V_{/\mathcal{R}_m} + V \times mV - V \times mSk\left(\overline{NG}\right)\Omega \tag{2.39}$$

The total kinetic energy W of the mechanical system is thus defined as:

$$W = \frac{1}{2}\begin{pmatrix} V_G \\ \Omega \end{pmatrix}^T \begin{pmatrix} P_d \\ \Pi \end{pmatrix} = \frac{1}{2}\begin{pmatrix} V \\ \Omega \end{pmatrix}^T \mathbf{M}_{RB}\begin{pmatrix} V \\ \Omega \end{pmatrix} \tag{2.40}$$

where:

- $\mathbf{M}_{RB} = \begin{pmatrix} mI_{3\times 3} & -mSk(NG) \\ mSk(NG) & I_N \end{pmatrix}$ is the lighter than air robot rigid body inertia matrix, m being the lighter than air robot total mass.
- $I_N = \begin{pmatrix} I_{xx} & 0 & -I_{xz} \\ 0 & I_{yy} & 0 \\ -I_{zx} & 0 & I_{zz} \end{pmatrix}$ is the moment of inertia, assuming the symmetry about the XZ plane.

Remark 2.8 The terms $\Omega \times mV - \Omega \times mSk(NG)\Omega$ are considered as Coriolis and centrifugal forces while the Coriolis and centrifugal moments terms are $\Omega \times mSk(NG)V + V \times mV - V \times mSk(NG)\Omega + \Omega \times I_N\Omega$.

Taking into account these assumptions, the analytic expressions of forces and moments acting on the lighter than air robot expressed in body-fixed frame $\mathcal{R}_m$, can be given by:

$$\mathbf{M}_{RB}\dot{\mathbf{V}} = \mathbf{C}(\mathbf{V})\mathbf{V} + \tau_s + \tau_a + \tau_p + \mathbf{M}_d\mathbf{V} \\ \dot{\eta} = \mathbf{RV} \\ \mathbf{V} = \begin{pmatrix} V \\ \Omega \end{pmatrix} \tag{2.41}$$

where:

- The Coriolis and centrifugal tensor $T_c = \mathbf{C}(\mathbf{V})\mathbf{V}$ is given by

$$T_c = \begin{pmatrix} \Omega \times mV - \Omega \times mSk(\overline{NG})\Omega \\ \Omega \times I_N\Omega + V \times mV + \Omega \times mSk(\overline{NG})V - V \times mSk(\overline{NG})\Omega \end{pmatrix}$$

where

$$\mathbf{C}(\mathbf{V}) = \begin{pmatrix} 0_{3\times 3} & -mSk(V + \Omega \times \overline{NG}) \\ -mSk(V + \Omega \times \overline{NG}) & -Sk(I_N\Omega) - mSk(NG)V \end{pmatrix}$$

is the Coriolis-centrifugal matrix associated to the lighter than air robot body mass, depending on its local velocities.

- τ_s is the static tensor due to the weight and lift forces.
- τ_p is the propulsion system tensor.
- τ_a is the aerodynamic tensor due to the atmosphere-lighter than air robot interaction and depends on the relative velocity.
- $\mathbf{M}_d = \begin{pmatrix} \dot{m}I_{3\times 3} & -\dot{m}Sk(\overline{NG}) \\ \dot{m}Sk(\overline{NG}) & \dot{I}_N \end{pmatrix}$

In presence of wind with velocity V_W, the aerodynamic forces and moment depend on the airship relative linear velocity ν_{1a} with respect to wind velocity, $V_a = V - V_W$, and hence $V_A = (V_a, \Omega)^T$.

2.3.2.2 Static Tensor

The gravity and lift forces must be considered in the following expressions:

$$F_{gravity} = R^T \begin{pmatrix} 0 \\ 0 \\ mg \end{pmatrix} \tag{2.42}$$

and

$$F_{lift} = -R^T \begin{pmatrix} 0 \\ 0 \\ B = \rho \cdot Vol \cdot g \end{pmatrix} \tag{2.43}$$

where B is the buoyancy force, ρ is the volumic mass of the inner gas, *Vol* is the lighter than air robot volume and g is the gravity acceleration.

The static force vector is given by the difference between the lighter than air robot weight (acting at the center of gravity G) and the lift force (acting upwards on the center of lift C). The static moment is given by the following relation

$$M_s = \overline{NG} \times F_{gravity} + \overline{NC} \times F_{lift} \tag{2.44}$$

where C represents the center of lift in the body fixed frame. Thus the static tensor can be given by:

$$\tau_s = \begin{pmatrix} F_{gravity} + F_{lift} \\ \overline{NG} \times F_{gravity} + \overline{NC} \times F_{lift} \end{pmatrix} \tag{2.45}$$

Remark 2.9 In hover mode, the lighter than air robot stability is mainly affected by its center of lift in relation to the center of gravity which can be adjusted to obtain either stable, neutral or unstable conditions. Putting all weight on the top would create a highly unstable lighter than air robot with a tendency to roll over in a stable position.

2.3.2.3 Propulsive Tensor

Actuators provide the means for maneuvering the lighter than air robot along its course, propelled by thrust. The required thrust and power change with flight conditions according to:

$$
\begin{aligned}
T_{req} &= \frac{1}{2}\rho(h)V^2 U_H^{2/3} C_{D0} \\
P_{req} &= \frac{T_{req}}{\upsilon}
\end{aligned}
\tag{2.46}
$$

where υ is the efficiency of the propellers, and $U_H^{2/3}$ is the standard aerodynamic reference area for airships.

Propellers are designed to exert thrust to drive the vehicle forward. The most popular propulsion system layout for pressurized lighter than air robot is twin ducted propellers mounted either side of the envelope bottom. Another one may exist in the stern for torque correction and attitude control. Hence, the lighter than air robot is an under-actuated system with two types of control: forces generated by thrusters and angular input μ_m controlling the direction of the thrusters. The lighter than air robot actuators list may be split into the following two sets:

1. Force inputs are available from the vectored two main propellers on each side of the gondola, providing a complementary lift to oppose the weighting mass, as well as a forward thrust controlling the longitudinal speed. When a differential input is added between two propellers, they also provide torque to control the rolling motion near hover. Finally a stern lateral thruster may be present to provide yaw control at low airspeed.
2. Surface deflections of the tail, when a minimum airspeed is present, provide torque inputs mostly for the control of the pitching and yawing motions.

When the air is perfectly still, the hover control is reduced to the use of the first set only.

Therefore for the first set, the propulsion torques can be written as:

$$
\tau_p = \begin{pmatrix} F_M \cos\mu \\ F_T \\ -F_M \sin\mu \\ 0 \\ F_M O_z \cos\mu + F_M O_x \sin\mu \\ F_T O_y \end{pmatrix}
\tag{2.47}
$$

where F_M, F_T represent respectively the main and tail torques, μ is the tilt angle of the main engine, the vector $\begin{pmatrix} O_x \\ 0 \\ O_z \end{pmatrix}$ is the position of the main engines while the vector $\begin{pmatrix} O_y \\ 0 \\ 0 \end{pmatrix}$ represent the position of the stern rotor. The surface deflections are considered in the aerodynamic tensor presented in the following paragraph.

Remark 2.10 Maximum in-flight incremental acceleration load factors owing to worst gusts or maneuvering, in general, may be: 0.75 g vertical, 0.5 g side and 0.5 g longitudinal.

2.3.2.4 Aerodynamic Tensor

The total aerodynamic forces and moments are related to

- Non stationary terms related to translational and rotational accelerations
- Terms involving translation-rotation products equivalent to Coriolis forces
- Terms involving rotation-rotation products equivalent to centrifugal forces
- Terms involving translation-translation products equivalent to stationary phenomena

Some elements of the theory of slender bodies are given in the following references [11, 12, 36, 64, 89, 108, 123, 207, 231].

Stationary Phenomena A pure translation depending aerodynamic tensor is considered. This phenomena arises from the forces and moment coming from the distribution of the pressure around the body and also the friction forces due to the viscosity of the air [123, 154, 207]. This tensor can expressed as:

$$\tau_{sta} = \frac{1}{2}\rho v_a^2 S_{ref} \begin{pmatrix} C_T \\ C_L \\ C_N \\ C_l \\ C_n \\ C_m \end{pmatrix} \tag{2.48}$$

where $C_L, C_N, C_T, C_l, C_m, C_n$ are aerodynamic coefficients depending on the geometrical shape of the airship and the positions of the control surfaces: rudders and elevators. These coefficients may be calculated in two different ways. The first one, is a pure experimental procedure which consists on collecting data in a wind tunnel, the second one, is an analytical procedure based on geometric quantities procedure [89, 113, 120, 131, 135, 137, 139, 149].

Remark 2.11 The bare hull of classic streamline form is directionally unstable, tending to turn broadside on the direction of motion. However, the ellipsoidal form remains a good compromise between structural and aerodynamic requirements for in-flight motion.

All aerodynamic lift and drag forces result from the combination of shear and pressure forces. Drag forces that are function of lift are known as induced drag.

Drag forces not strongly related to lift are usually known as parasite drag or zero lift drag [63].

If the theoretical pressure forces in a perfect fluid are integrated over a streamlined body without flow separation, the pressure over a body that yields a drag force in the flight direction should be exactly matched by the pressure around the body which yield a forward force. Thus, if skin friction is ignored, the net drag would be zero. This is known to be false, because even if the lift of a body of revolution is less than the lift due to a wing, it is different from zero.

The boundary layer which is produced by viscosity causes the flow to separate somewhere on the back half of the lighter than air robot. This prevents the full attainment of the forward acting force leaving a net drag force due to viscous separation. This drag force depends upon the location of the separation point of the body. This separation point is function of the curvature of the body and the amount of energy in the flow. Thus, Kirchoff's theory must be completed by the theory of slender bodies, initiated by Munk (1924) [154]. This theory is expressed for bodies whose thickness ratio $\ll 1$ (ratio between the maximum diameter of the hull and its length). Typically airship hulls experience skin friction drag, form (or pressure) drag and induced drag.

Owing to the large surface area of the airship, skin friction is the largest fraction of the total drag. For this reason, the reference area S_{ref} (the wetted area) is in general approximated as the surface area. However, another approach is to consider the surface area as $(Buoyant \cdot Volum)^{2/3}$.

In the stationary part, two additive terms appear [89, 208, 217]:

– The first one for low values of the angle of attack α

$$\begin{aligned} C_{L1} &= C_1 \cos\alpha \sin 2\beta \\ C_{N1} &= C_1 \sin 2\alpha \cos^2\beta \\ C_{n1} &= C_0 \cos\alpha \sin 2\beta \\ C_{m1} &= C_2 \sin 2\alpha \cos^2\beta \end{aligned} \tag{2.49}$$

β being the side-slip angle

– The second one in turbulent situation, augmenting rapidly with the angle of attack

$$\begin{aligned} C_{L2} &= C_3 \sin\beta \sqrt{\sin^2\alpha \cos^2\beta + \sin^2\beta} \\ C_{N2} &= C_3 \sin\alpha \cos\beta \sqrt{\sin^2\alpha \cos^2\beta + \sin^2\beta} \\ C_{l2} &= (C_4 \sin\alpha + C_{10}\cos\alpha)\sin 2\beta \\ C_{m2} &= C_5 \sin\alpha \cos\beta \sqrt{\sin^2\alpha \cos^2\beta + \sin^2\beta} \\ C_{n2} &= C_9 \sin\beta \sqrt{\cos^2\alpha \cos^2\beta + \sin^2\beta} \end{aligned} \tag{2.50}$$

Finally, the following general relations are obtained:

$$\left(\begin{array}{c} C_L = C_{L1} + C_{L2} \\ C_N = C_{N1} + C_{N2} \\ C_T = \left[C_6 \left(\mathrm{sign}(C_L)\sqrt{C_N^2 + C_L^2} + C_7 \right)^2 + C_8 \right] \cos^2\alpha \cos^2\beta \\ C_l = C_{l2} \\ C_n = C_{n1} + C_{n2} \\ C_m = C_{m1} + C_{m2} \end{array}\right) \tag{2.51}$$

where the C_i, $i = 0..10$ are constants depending on the reference area, the separation point location, the base surface at this point and the diameter function of the lighter than air robot.

Non Stationary Aerodynamic Phenomena When the lighter than air robot moves, the air close to its body is moved. Contrary to the other aerial vehicles, the mass of displaced air is close to those of the lighter than air robot and consequently cannot be neglected. The displaced air mass is known as *added mass* or *virtual mass*. Added mass and inertia coefficients received their name because they can be linearly combined with the true lighter than air robot mass in the equations of motion to form one coefficient [11, 36, 42, 64, 80, 137, 207].

Each component of the airship is represented by an ellipsoid with three independently sized principal axes, this allows the added masses to be calculated analytically. Ellipsoid geometry, orientation and relative location are chosen so that both added masses and added moments of inertia are optimally modeled. Interference effects between the main hull component and an appendage are approximately accounted for by using the flow field around a replacement ellipsoid for the hull to modify the flow at the appendage, interference effects between appendages are neglected. The analysis uses incompressible potential flow theory. It does not account for any circulation in the flow [214].

The added mass matrix is, in general, an extra-diagonal matrix:

$$\mathbf{M}_A = -\begin{pmatrix} X_{\dot u} & X_{\dot v} & X_{\dot w} & X_{\dot p} & X_{\dot q} & X_{\dot r} \\ Y_{\dot u} & Y_{\dot v} & Y_{\dot w} & Y_{\dot p} & Y_{\dot q} & Y_{\dot r} \\ Z_{\dot u} & Z_{\dot v} & Z_{\dot w} & Z_{\dot p} & Z_{\dot q} & Z_{\dot r} \\ K_{\dot u} & K_{\dot v} & K_{\dot w} & K_{\dot p} & K_{\dot q} & K_{\dot r} \\ M_{\dot u} & M_{\dot v} & M_{\dot w} & M_{\dot p} & M_{\dot q} & M_{\dot r} \\ N_{\dot u} & N_{\dot v} & N_{\dot w} & N_{\dot p} & N_{\dot q} & N_{\dot r} \end{pmatrix} \tag{2.52}$$

The inertial effects of this added mass constitute the first component of the aerodynamic tensor. Another part of the aerodynamic forces is coming from the translation-rotation and rotation-rotation coupling motions and can be assimilated to Coriolis-centrifugal effects associated to the added mass and can also be represented as a damping effect representation. Due to the importance of the added mass of the lighter than air robot, this tensor must be included. In addition, a pure translation depending aerodynamic tensor is considered. The total added masses of the vehicle are taken to be the sum of the added masses of each of the component ellipsoids, except that interference effects between the hull and fins are accounted for first. The interference effects can be modeled approximately by superimposing a uniform velocity over a relatively small ellipsoid which finds itself in the presence of a larger one; this interference velocity is the velocity field around the large ellipsoid evaluated at the location of the centroid of the smaller one, without the smaller ellipsoid being present.

The added mass matrix of a rigid body lighter than air robot must include the contributions of both the hull and the fins as:

$$M_{A_{rigid}} = M_{AH} + M_{AF} \tag{2.53}$$

It is in general acknowledged that added mass coefficients are not strongly dependent on viscous effects (such as circulation and boundary layer growth). Indeed, many of the experimental techniques currently in use for determining the coefficients make this assumption since they ignore the dependence of added mass on both the orientation of the vehicle to the oncoming flow and time in an unsteady flow. A potential flow analysis, neglecting circulation, is an appropriate method for estimating the added mass coefficient.

Fortunately, many of the added mass derivatives contained in the general expressions for added mass are either zero or mutually related when the body has various symmetries.

Added Mass Derivatives for a Prolate Ellipsoid An ellipsoid considered with the origin at the center of the ellipsoid, can be described as:

$$\frac{x^2}{a^2} + \frac{y^2}{b^2} + \frac{z^2}{c^2} = 1 \tag{2.54}$$

Here a, b, c are the semi-axes. A prolate spheroid is obtained by letting $b = c$ and $a > b$. Cross coupling terms will be zero due to body symmetry about three planes:

$$\begin{gathered} X_{\dot{u}} = -m\frac{\alpha_0}{2-\alpha_0} \\ Y_{\dot{v}} = Z_{\dot{w}} = -m\frac{\beta_0}{2-\beta_0} \\ K_{\dot{p}} = 0 \\ N_{\dot{r}} = M_{\dot{q}} = -\frac{1}{5}\frac{(b^2-a^2)^2(\alpha_0-\beta_0)}{2(b^2-a^2)-(b^2+a^2)(\alpha_0-\beta_0)} \end{gathered} \tag{2.55}$$

where the mass of the prolate spheroid is

$$m = \frac{4}{3}\pi\rho a b^2 \tag{2.56}$$

The eccentricity e is introduced

$$e^2 = 1 - \frac{b^2}{a^2} \tag{2.57}$$

Hence, the constants α_0 and β_0 can be calculated as

$$\begin{aligned}\alpha_0 &= \frac{2(1-e^2)}{e^3}\left(\frac{1}{2}\ln\left(\frac{1+e}{1-e}\right) - e\right)\\ \beta_0 &= \frac{1}{e^2} - \left(\frac{1-e^2}{2e^3}\ln\left(\frac{1+e}{1-e}\right)\right)\end{aligned} \tag{2.58}$$

An alternative representation of these mass derivatives is presented by Lamb [123] who defines Lamb's k-factors as

$$\begin{aligned}k_1 &= \frac{\alpha_0}{2-\alpha_0}\\ k_2 &= \frac{\beta_0}{2-\beta_0}\\ k' &= \frac{e^4(\beta_0-\alpha_0)}{(2-e^2)(2e^2-(2-e^2)(\beta_0-\alpha_0))}\end{aligned} \tag{2.59}$$

Hence the definition of the added mass derivatives simplifies to

$$\begin{aligned}X_{\dot{u}} &= -k_1 m\\ Y_{\dot{v}} = Z_{\dot{w}} &= -k_2 m\\ N_{\dot{r}} = M_{\dot{q}} &= -k' I_y\end{aligned} \tag{2.60}$$

where the moment of inertia of the prolate spheroid is

$$I_y = I_z = \frac{4}{15}\pi\rho a b^2(a^2+b^2) \tag{2.61}$$

The lighter than air robot hull is generally responsible for the greatest contribution to the added mass coefficient. In practice, a simple approach to obtain the added mass and moments of inertia of the hull is to approximate the hull as a set of ellipsoids of revolution. For each ellipsoid, all the off-diagonal terms in the added mass matrix are considered to be zero.

Fins added masses are calculated semi-empirically using flat plat models. The added mass and moment of inertia of the fins can be computed by integrating the 2D added mass of the cross section over the fin region. The contribution of the fins to these 2D added mass terms can be written as:

$$\begin{aligned}\overline{m}_{F,22} = \overline{m}_{F,33} &= \rho\pi\left(b - \frac{R_s^2}{b}\right)^2\\ \overline{m}_{F,44} &= \frac{2}{\pi}k_{44}\rho b^4\end{aligned} \tag{2.62}$$

where R_s is the hull cross sectional radius and b is the fin semi span. The factor k_{44} is a function of R_s/b. The non-zero elements in the added mass matrix of the fins are obtained from the following integrals:

$$m_{F,22} = m_{F,33} = \eta_f \int_{x_{FS}}^{x_{FE}} \left(\overline{m}_{F,22} dx\right) \tag{2.63}$$

$$m_{F,35} = -m_{F,26} = m_{F,33} = \eta_f \int_{x_{FS}}^{x_{FE}} \left(\overline{m}_{F,22} x dx\right) \tag{2.64}$$

$$m_{F,44} = \eta_f \int_{x_{FS}}^{x_{FE}} \left(\overline{m}_{F,44} dx\right) \tag{2.65}$$

$$m_{F,55} = m_{F,66} = \eta_f \int_{x_{FS}}^{x_{FE}} \left(\overline{m}_{F,22} x^2 dx\right) \tag{2.66}$$

where x_{FS}, x_{FE} are respectively the x coordinates of the start and end positions of the fins. An efficiency factor η_f is included to account for 3D effects.

Complete Lighter than Air Robot Model To make an operational lighter than air robot, it is necessary to provide control surfaces and propulsion units (in the gondola). Each of these items has its own drag modified when it is fitted to the hull as it is then operating in the flow field of the hull and vice-versa. The ailerons modify all the forces and moments coefficients. However, correctly dimensioned and situated, they stabilize the streamlined body. The gondola introduces an asymmetry in the geometry of the vehicle. It influences all the forces sometimes only modifying the first and second term introduced above through the coefficients C_i and sometimes adding a third term independent of the angle of attack α and the skid angle β.

Hence, the aerodynamical tensor, defined by the contribution of all aerodynamic phenomena, can be synthesized as: Coriolis-centrifugal effects associated to the added mass and can also be represented as an damping effect representation, i.e. $D(\Omega)V_A$. Moreover, the pure translation motion action tensor, $\tau_1 = (0_{1\times 3}, (V \times M_{A_{rigid}} V)^T)^T$, is usually ignored in the other aerial applications.

Using the Complementarity between the coefficients of Kirchoff and Bryson theories, the following damping coefficient $\mathbf{D} = [D_V \ D_\Omega]^T$, is obtained:

$$D_V = \begin{pmatrix} 0 & a_{22} r & -a_{33} q \\ p m_{13} + r(x_{m_{11}} - a_{11}) & 0 & a_{33} p \\ (a_{11} - x_{m_{22}}) q & -a_{22} p & 0 \\ p m_{33} + r(a_{15} + x_{m_{13}}) & -(a_{62} + a_{35}) q & (a_{62} + a_{35}) r + a_{24} p \\ (a_{35} + x_{m_{33}}) q & -a_{42} r + a_{62} p & -a_{15} q \\ -(a_{51} + a_{24} - x_{m_{13}}) p - (a_{26} - x_2 m_{11}) r & (a_{15} + a_{42}) q & -a_{53} p \end{pmatrix}$$

$$D_\Omega = \begin{pmatrix} a_{24}r & -a_{35}q & a_{26}r \\ a_{35}q & -a_{15}q & 0 \\ -a_{24}p - a_{26}r & a_{15}q & 0 \\ -a_{64}q & (a_{55}-a_{66})r & 0 \\ a_{64}p + (a_{66}-a_{44})r & 0 & -a_{64}r \\ (a_{44}-a_{55})q & a_{46}r & 0 \end{pmatrix} \tag{2.67}$$

The aerodynamic torques are thus given by the following relation:

$$\tau_a = \mathbf{M}_{RB}\dot{\mathbf{V}}_A + \mathbf{D}(\Omega)\mathbf{V}_A + \tau_{sta} \tag{2.68}$$

2.3.2.5 Complete Model with Euler Angles

The complete mechanical-aerodynamical lighter than air robot model can be given by:

$$\mathbf{M}\dot{\mathbf{V}} = \mathbf{M}_{RB}\dot{\mathbf{V}}_w + C(\mathbf{V})\mathbf{V} + \tau_s + \mathbf{D}(\Omega)\mathbf{V}_A + \tau_{sta} + \tau_p + \mathbf{M}_d\mathbf{V} \tag{2.69}$$

$$\dot{\eta} = \mathbf{R}\mathbf{V} \tag{2.70}$$

where $\mathbf{M} = \mathbf{M}_{A_{rigid}} + \mathbf{M}_{RB}$ is the total mass matrix.

Remark 2.12 For a system with added masses, the term $V \times MV$ is different from zero. The terms $V \times MV$, $\Omega \times MV$, $V \times M\Omega$ and $\Omega \times M\Omega$ show the centrifugal and Coriolis components.

Another modeling approach is considered in the next section.

2.3.3 6 DOF Dynamics: Lagrange Approach

One advantage with the Lagrangian approach is to deal with two scalar energy functions $\mathbf{W}$ (kinetic energy) and E_p (potential energy). The Newtonian approach is vector oriented since everything is derived from Newton's second law. The Lagrangian description uses generalized coordinates (one for each degree of freedom), all of which must be independent. The equations of motion for lighter than air vehicles are derived in a Lagrangian framework. The Lagrangian approach has several distinctive features to the Newtonian approach. The derivation of the aerodynamic added inertia and the vehicle's rigid body equations of motion can be done in a common framework [123, 143, 150, 231]. The added inertia is given a physical interpretation when the vehicle-ambient air system is considered from an energy point of view instead of a force moment approach. Recall that:

$$\frac{d}{dt}\left(\frac{\partial L}{\partial \dot{\eta}}\right) - \frac{\partial L}{\partial \eta} + \frac{\partial P_d}{\partial \dot{\eta}} = \tau_\eta = \sum_{i=1}^{m} F_i(\eta)u_i \tag{2.71}$$

where L is the Lagrangian F defines the input directions and the additional term describes the dissipative forces P_d as a power function.

$$\frac{\partial P_d}{\partial \dot{\eta}} = D_\eta\left(\eta, \nu\right)\dot{\eta} \tag{2.72}$$

Forces dissipating energy are called dissipative forces: friction, air drag...The generalized coordinates are taken to be $\eta = (x, y, z, \phi, \theta, \psi)^T$ so that $\mathbf{V} = (u, v, w, p, q, r)^T$ are dependent variables related to the generalized coordinates [11, 12, 41, 42, 64, 105, 207].

A rigid body moving in an unbounded fluid is holonomic due to, the unboundedness and the infinite extent of the fluid. Hence a rigid body moving without constraints in 6 degrees of freedom is an ordinary Lagrangian system. However, the infinite degrees of freedom system by the ambient air particles cannot be said to be holonomic.

The Lagrangian for the vehicle-ambient air system is defined as:

$$L = W_{RB} + W_A - E_p \tag{2.73}$$

where W_{RB} is the rigid body kinetic energy, W_A is the fluid kinetic energy and E_p is the potential energy defined implicitly by:

$$\frac{\partial E_p}{\partial \eta} = \tau_s(\eta) \tag{2.74}$$

Hence the total kinetic energy can be expressed as:

$$W_{RB} + W_A = \frac{1}{2}\dot{\eta}^T\mathbf{R}^{-T}\left(M_{RB} + M_{A_{rigid}}\right)\mathbf{R}^{-1}\dot{\eta} \tag{2.75}$$

where the following notation is used: $\mathbf{R}^{-T} = (\mathbf{R}^{-1})^T$.

The equations of motion for lighter than air vehicles can be written as follows:

$$\begin{gathered}\mathbf{M}\dot{\mathbf{V}} + \mathbf{C}(\mathbf{V})\mathbf{V} + \mathbf{D}(\mathbf{V})\mathbf{V} + \tau_s(\eta) = \varpi(\phi) + \tau_a + \tau_p \\ \dot{\eta} = \mathbf{R}\mathbf{V}\end{gathered} \tag{2.76}$$

$\mathbf{M}$ is the inertia matrix including the aerodynamic virtual inertia (added mass), $\mathbf{C}(\mathbf{V})\mathbf{V}$ contains the nonlinear forces and moments due to centrifugal and Coriolis forces and $\mathbf{D}(\mathbf{V})\mathbf{V}$ is the vehicle damping matrix where the potential damping and the viscous effects are lumped together. $\tau_s(\eta)$ is a vector containing the restoring terms formed by the vehicle's buoyancy and gravitational terms , $\varpi(\phi)$ is the atmospheric disturbance vector and $\tau_a + \tau_p$ is a tensor containing the aerodynamic and propulsion tensors. $\mathbf{R}$ is the velocity transformation matrix which transforms velocities from the body fixed frame $\mathcal{R}_m$ to the Earth fixed reference frame $\mathcal{R}_f$.

2.3.3.1 Kinetic Energy

The kinetic energy of a solid in motion can be written as:

$$W_{RB}(\eta,\dot{\eta}) = \frac{1}{2}\int\int\int\left[\rho\left(V + (\Omega \times r)\right)^T \left(V + (\Omega \times r)\right)\right] dv \tag{2.77}$$

or equivalently:

$$W_{RB}(\eta,\dot{\eta}) = \frac{1}{2}\dot{\eta}^T \mathbf{M}_{RB}\dot{\eta} \tag{2.78}$$

with the symmetric positive definite matrix:

$$\mathbf{M}_{RB} = \begin{pmatrix} m & 0 & 0 & 0 & mz_g & -my_g \\ 0 & m & 0 & -mz_g & 0 & mx_g \\ 0 & 0 & m & my_g & -mx_g & 0 \\ 0 & -mz_g & my_g & I_x & -I_{xy} & -I_{xz} \\ mz_g & 0 & -mx_g & -I_{xy} & I_y & -I_{yz} \\ -my_g & mx_g & 0 & -I_{xz} & -I_{yz} & I_z \end{pmatrix} \tag{2.79}$$

The radiation induced forces and moments can identified as the sum of three components:

- Added mass due to the inertia of the surrounding fluid
- Radiation induced potential damping due to the energy carried away by the wind
- Weight and buoyancy (Archimedes) forces

Added mass should be understood as pressure-induced forces and moments due to a forced harmonic motion of the body which are proportional to the acceleration of the body. In order to allow the vehicle to pass through the air, the fluid must move aside and then close behind the vehicle. As a consequence, the fluid passage possesses kinetic that it would lack if the vehicle was not in motion. The kinetic energy of the air particles generated by a vehicle in motion can be written as:

$$W_A(\eta,\dot{\eta}) = \frac{1}{2}\dot{\eta}^T \mathbf{M}_{A_{rigid}}\dot{\eta} \tag{2.80}$$

with $\mathbf{M}_{A_{rigid}}$ the added mass and inertia matrix, symmetric positive definite. Mathematically, added mass is the proportionality constant relating the kinetic energy in the air surrounding the vehicle to the square of the vehicle speed, in the same way that vehicle mass relates vehicle kinetic energy to speed squared. During vehicle accelerations, both the kinetic energy of the vehicle and the air are changed, their ratio remaining constant in an ideal fluid. Therefore, an accelerating vehicle must overcome the effective inertia of both the vehicle and added mass. The added mass for a lighter than air robot can be equivalent to the actual vehicle mass, and so must be accounted for properly.

Remark 2.13 The air is assumed to be an incompressible and homogeneous fluid of infinite extent.

2.3.3.2 Lighter than Air Robot Model

Let the motion of the lighter than air robot be described by its inertial velocity $\mathbf{V}$ a 6D vector including the inertial linear V and angular Ω velocities. Let the surrounding air be described by an inertial wind velocity $\mathbf{V}_W = (V_W, \Omega_W)^T$. The lighter than air robot has thus a relative air velocity $\mathbf{V}_a$:

$$\mathbf{V}_a = \mathbf{V} - \mathbf{V}_W \tag{2.81}$$

The total kinetic energy is defined as a sum [9, 80, 150, 206, 214]:

$$\mathbf{W} = \mathbf{W}_G + \mathbf{W}_B + \mathbf{W}_V \tag{2.82}$$

Accounting for:

- $\mathbf{W}_G$: The kinetic energy of the vehicle expressed in the center of gravity G located at a distance $\overline{NG}$ from the body fixed frame center N:
- $\mathbf{W}_B$: The kinetic energy added to the buoyancy air (displaced by the lighter than air robot volume) expressed in the center of lift C and where $\mathbf{M}_B$ is the generalized mass matrix of the buoyancy air. In the case of the buoyancy air, the kinetic energy is the difference between the extra kinetic energy and the contribution of the air mass motion.
- $\mathbf{W}_V$: The energy due to an extra virtual mass also expressed in C and where $\mathbf{M}_v$ is the generalized virtual mass matrix of the mass of air around the lighter than air robot and displaced with its relative motion in the air.

where

$$\mathbf{W}_G = \frac{1}{2}\mathbf{V}_G^T \mathbf{M}_G \mathbf{V}_G \tag{2.83}$$

with

$$\mathbf{W}_B = \frac{1}{2}\mathbf{V}_a^T \mathbf{M}_B \mathbf{V}_a - \frac{1}{2}\mathbf{V}^T \mathbf{M}_B \mathbf{V} \tag{2.84}$$

and

$$\mathbf{W}_v = \frac{1}{2}\mathbf{V}_a^T \mathbf{M}_v \mathbf{V}_a \tag{2.85}$$

where the generalized mass matrices being given by

$$\mathbf{M}_G = \begin{pmatrix} mI_{3\times 3} & 0 \\ 0 & I_G \end{pmatrix} \qquad \mathbf{M}_B = \begin{pmatrix} m_B I_{3\times 3} & 0 \\ 0 & J_B \end{pmatrix} \qquad \mathbf{M}_v = \begin{pmatrix} M_v & 0 \\ 0 & I_v \end{pmatrix} \tag{2.86}$$

Therefore, the total kinetic energy may be written as:

$$\mathbf{W} = \frac{1}{2}\mathbf{V}_G^T \mathbf{M}_G \mathbf{V}_G + \frac{1}{2}\mathbf{V}_a^T \mathbf{M}_v \mathbf{V}_a - \frac{1}{2}\mathbf{V}^T \mathbf{M}_B \mathbf{V} + \frac{1}{2}\mathbf{V}_a^T \mathbf{M}_B \mathbf{V}_a \tag{2.87}$$

All terms of the kinetic energy in the body-fixed frame consider that the linear speed of the center of gravity G is related to the linear speed of the body fixed frame center: the bow N, through the angular speed:

$$\mathbf{V}_G = \begin{pmatrix} I_{3\times 3} & -Sk(\overline{NG}) \\ 0_{3\times 3} & I_{3\times 3} \end{pmatrix} \mathbf{V} \tag{2.88}$$

Substituting in the previous Eq. (2.87) gives:

$$\mathbf{W} = \frac{1}{2}\mathbf{V}^T\mathbf{M}_N\mathbf{V} - \frac{1}{2}\mathbf{V}^T\mathbf{M}_B\mathbf{V} + \frac{1}{2}\mathbf{V}_a^T\mathbf{M}_B\mathbf{V}_a + \frac{1}{2}\mathbf{V}_a^T\mathbf{M}_v\mathbf{V}_a \tag{2.89}$$

with

$$\mathbf{M}_N = \begin{pmatrix} mI_{3\times 3} & -mSk(\overline{NG}) \\ mSk(\overline{NG}) & I_G - m(Sk(\overline{NG}))^2 \end{pmatrix} \tag{2.90}$$

Therefore,

$$\mathbf{W} = \frac{1}{2}(\mathbf{V} - \mathbf{V}_W)^T(\mathbf{M}_v + \mathbf{M}_B)(\mathbf{V} - \mathbf{V_W}) + \frac{1}{2}\mathbf{V}^T(\mathbf{M}_N - \mathbf{M_B})\mathbf{V} \tag{2.91}$$

or

$$\mathbf{W} = \frac{1}{2}\mathbf{V}^T\left(\mathbf{M}_N + \mathbf{M}_v\right)\mathbf{V} + \frac{1}{2}\mathbf{V}_w^T\left(\mathbf{M}_B + \mathbf{M}_v\right)\mathbf{V}_w - \mathbf{V}^T\left(\mathbf{M}_B + \mathbf{M}_v\right)\mathbf{V}_w \tag{2.92}$$

Finally, the kinetic energy can be expressed as a function of the lighter than air robot and wind inertial velocities

$$\mathbf{W} = \mathbf{W}_1 + \mathbf{W}_2 + \mathbf{W}_3 \tag{2.93}$$

or equivalently:

$$\mathbf{W}_1 = \frac{1}{2}\mathbf{V}^T\mathbf{M}_{vn}\mathbf{V} \tag{2.94}$$

with

$$\mathbf{W}_2 = \frac{1}{2}\mathbf{V}_W^T\mathbf{M}_{Bv}\mathbf{V}_W \tag{2.95}$$

and

$$\mathbf{W}_3 = -\mathbf{V}^T\mathbf{M}_{Bv}\mathbf{V}_W \tag{2.96}$$

where

$$\mathbf{M}_{vn} = \mathbf{M}_N + \mathbf{M}_v = \begin{pmatrix} mI_{3\times 3} + M_v & -mSk(\overline{NG}) \\ mSk(\overline{NG}) & I_G - m(Sk(\overline{NG}))^2 + I_v \end{pmatrix}$$

$$\mathbf{M}_{Bv} = \mathbf{M}_v + \mathbf{M}_B = \begin{pmatrix} m_B I_{3\times 3} + M_v & 0 \\ 0 & J_B + I_v \end{pmatrix}$$

Therefore, the Lagrange-Hamilton equations of motion are applied to each of the three terms of the kinetic energy in the sequel.

2.3.3.3 First Term of the Kinetic Energy

The first term of the kinetic energy corresponding to the no wind case is then:

$$\mathbf{W}_1 = \frac{1}{2}\mathbf{V}^T\mathbf{M}_{vn}\mathbf{V} = \frac{1}{2}\dot{\eta}^T\mathbf{R}^{-T}\mathbf{M}_{vn}\mathbf{R}^{-1}\dot{\eta} \tag{2.97}$$

The partial derivative of this kinetic energy $\mathbf{W}_1$ relative to $\dot{\eta}$ is

$$\frac{\partial \mathbf{W}_1}{\partial \dot{\eta}} = \mathbf{R}^{-T}\mathbf{M}_{vn}\mathbf{R}^{-1}\dot{\eta} \tag{2.98}$$

since

$$\mathbf{M}_{vn}^T = \mathbf{M}_{vn} \tag{2.99}$$

Now taking the time derivative gives:

$$\begin{aligned}\frac{d}{dt}\left(\frac{\partial \mathbf{W}_1}{\partial \dot{\eta}}\right) &= \dot{\mathbf{R}}^{-T}\mathbf{M}_{vn}\mathbf{R}^{-1}\dot{\eta} + \mathbf{R}^{-T}\dot{\mathbf{M}}_{vn}\mathbf{R}^{-1}\dot{\eta}\\ &\quad + \mathbf{R}^{-T}\mathbf{M}_{vn}\dot{\mathbf{R}}^{-1}\dot{\eta} + \mathbf{R}^{-T}\mathbf{M}_{vn}\mathbf{R}^{-1}\ddot{\eta}\end{aligned} \tag{2.100}$$

Or equivalently:

$$\frac{d}{dt}\left(\frac{\partial \mathbf{W}_1}{\partial \dot{\eta}}\right) = \dot{\mathbf{R}}^{-T}\mathbf{M}_{vn}\mathbf{V} + \mathbf{R}^{-T}\dot{\mathbf{M}}_{vn}\mathbf{V} + \mathbf{R}^{-T}\mathbf{M}_{vn}\dot{\mathbf{V}} \tag{2.101}$$

Where the following relations are used:

$$\begin{aligned}\dot{\mathbf{R}}^{-1}\mathbf{R} + \mathbf{R}^{-1}\dot{\mathbf{R}} &= \mathbf{0}_{6\times 6}\\ \dot{\mathbf{V}} &= \mathbf{R}^{-1}\ddot{\eta} - \mathbf{R}^{-1}\dot{\mathbf{R}}\mathbf{R}^{-1}\dot{\eta}\end{aligned} \tag{2.102}$$

The partial derivative of the kinetic energy $\mathbf{W}_1$ relative to η is thus:

$$\frac{\partial \mathbf{W}_1}{\partial \eta} = \frac{1}{2}\frac{\partial}{\partial \eta}\left(\dot{\eta}^T\mathbf{R}^{-T}\mathbf{M}_{vn}\mathbf{R}^{-1}\dot{\eta}\right) = \mathbf{K}\mathbf{M}_{vn}\mathbf{V} \tag{2.103}$$

with

$$\mathbf{K} = \frac{\partial}{\partial \eta}\left(\mathbf{R}^{-1}\dot{\eta}\right) \tag{2.104}$$

or

$$\begin{pmatrix}\frac{\partial}{\partial \eta_1}\left(\mathbf{R}^{-1}\dot{\eta}\right)\\ \frac{\partial}{\partial \eta_2}\left(\mathbf{R}^{-1}\dot{\eta}\right)\end{pmatrix} = \begin{pmatrix}0_{3\times 3} & 0_{3\times 3}\\ K_1 & K_2\end{pmatrix} \tag{2.105}$$

The generalized force relative to the kinetic energy with no wind is then obtained:

$$\mathbf{F}_1(\eta,\dot{\eta}) = \mathbf{R}^{-T}\mathbf{M}_{vn}\dot{\mathbf{V}} + \left(\mathbf{R}^{-T}\dot{\mathbf{M}}_{vn} + \dot{\mathbf{R}}^{-T}\mathbf{M}_{vn} - \mathbf{K}\mathbf{M}_{vn}\right)\mathbf{V} \tag{2.106}$$

2.3.3.4 Second Term of the Kinetic Energy

The second term of the kinetic energy is computed, defining the wind coordinates vector as $\eta_W = (x_W, y_W, z_W, \phi_W, \theta_W, \psi_W)^T$ whose time derivative is related to the wind velocity:

$$\dot{\eta}_W = \mathbf{R}\mathbf{V}_W \tag{2.107}$$

The second term of the kinetic energy corresponds to:

$$\mathbf{W}_2 = \frac{1}{2}\mathbf{V}_W^T\mathbf{M}_{Bv}\mathbf{V}_W = \frac{1}{2}\dot{\eta}_W^T\mathbf{R}^{-T}\mathbf{M}_{Bv}\mathbf{R}^{-1}\dot{\eta}_W \tag{2.108}$$

with

$$\frac{d}{dt}\left(\frac{\partial\mathbf{W}_2}{\partial\dot{\eta}}\right) = \frac{d}{dt}\left(\frac{\partial}{\partial\dot{\eta}}\left(\frac{1}{2}\dot{\eta}_W^T\mathbf{R}^{-T}\mathbf{M}_{Bv}\mathbf{R}^{-1}\dot{\eta}_W\right)\right) = 0 \tag{2.109}$$

The partial derivative of the kinetic energy $\mathbf{W}_2$ relative to the generalized coordinate η is

$$\frac{\partial\mathbf{W}_2}{\partial\eta} = \frac{\partial}{\partial\eta}\left(\frac{1}{2}\dot{\eta}_W^T\mathbf{R}^{-T}\mathbf{M}_{Bv}\mathbf{R}^{-1}\dot{\eta}_W\right) \tag{2.110}$$

or

$$\frac{\partial\mathbf{W}_2}{\partial\eta} = \mathbf{K}_W\mathbf{M}_{Bv}\mathbf{V}_W \tag{2.111}$$

where

$$\mathbf{K}_W = \begin{pmatrix} 0_{3\times3} & 0_{3\times3} \\ \mathbf{K}_{W_1} & \mathbf{K}_{W_2} \end{pmatrix} \tag{2.112}$$

The generalized force relative to is then obtained:

$$\mathbf{F}_2 = -\mathbf{K}_W\mathbf{M}_{Bv}\mathbf{V}_W \tag{2.113}$$

2.3.3.5 Third Term of the Kinetic Energy

Applying the same procedure to the third term, $W_3 = -\dot{\eta}^T\mathbf{R}^{-T}\mathbf{M}_{Bv}\mathbf{V}_W$ we obtain:

$$\frac{d}{dt}\left(\frac{\partial\mathbf{W}_3}{\partial\dot{\eta}}\right) = -\dot{\mathbf{R}}^{-T}\mathbf{M}_{Bv}\mathbf{V}_W - \mathbf{R}^{-T}\mathbf{M}_{Bv}\dot{\mathbf{V}}_W \tag{2.114}$$

leading to:

$$\mathbf{F}_3(\eta,\dot{\eta}) = -\mathbf{R}^{-T}\mathbf{M}_{Bv}\dot{\mathbf{V}}_W - \dot{\mathbf{R}}^{-T}\mathbf{M}_{Bv}\mathbf{V}_W + \mathbf{K}\mathbf{M}_{Bv}\mathbf{V}_W + \mathbf{K}_W\mathbf{M}_{Bv}\mathbf{V} \tag{2.115}$$

2.3.3.6 Forces

Summing up all the generalized forces, we have:

$$\mathbf{F}(\eta,\dot{\eta}) = \mathbf{R}^{-T}\mathbf{M}_{vn}\dot{\mathbf{V}} + \left(\mathbf{R}^{-T}\dot{\mathbf{M}}_{vn} + \dot{\mathbf{R}}^{-T}\mathbf{M}_{vn} - \mathbf{K}\mathbf{M}_{vn}\right)\mathbf{V} - \mathbf{K}_W\mathbf{M}_{Bv}\mathbf{V}_W$$
$$- \mathbf{R}^{-T}\mathbf{M}_{Bv}\dot{\mathbf{V}}_W - \dot{\mathbf{R}}^{-T}\mathbf{M}_{Bv}\mathbf{V}_W + \mathbf{K}\mathbf{M}_{Bv}\mathbf{V}_W + \mathbf{K}_W\mathbf{M}_{Bv}\mathbf{V} \quad (2.116)$$

Let's introduce

$$\mathbf{F}(\mathbf{V}) = \mathbf{R}^T\mathbf{F}(\eta,\dot{\eta}) \quad (2.117)$$

Thus

$$\mathbf{F}(\mathbf{V}) = \mathbf{M}_{vn}\dot{\mathbf{V}} + (\Omega_6 + \mathbf{V}_6)\mathbf{M}_{vn}\mathbf{V} + \dot{\mathbf{M}}_{vn}\mathbf{V} - \mathbf{M}_{Bv}\dot{\mathbf{V}}_W$$
$$- (\Omega_6 + \mathbf{V}_6)\mathbf{M}_{Bv}\mathbf{V}_W - \mathbf{V}_{W_6}\mathbf{M}_{Bv}\mathbf{V} \quad (2.118)$$

where

$$\mathbf{V}_6 = \begin{pmatrix} 0_{3\times 3} & 0_{3\times 3} \\ Sk(V) & 0_{3\times 3} \end{pmatrix} \quad (2.119)$$

and

$$\Omega_6 = \begin{pmatrix} Sk(\Omega) & 0_{3\times 3} \\ 0_{3\times 3} & Sk(\Omega) \end{pmatrix} \quad (2.120)$$

and

$$\mathbf{V}_{w_6} = \begin{pmatrix} 0_{3\times 3} & 0_{3\times 3} \\ Sk(V_W) & 0_{3\times 3} \end{pmatrix} \quad (2.121)$$

where $Sk(V)$, $Sk(\Omega)$, $Sk(V_W)$ denote the antisymmetric cross product matrices corresponding to the vectors V, Ω, V_W and using the following equations

$$Sk(V) = -J^T\mathbf{K}_1$$
$$Sk(\Omega) = J^T\left((\dot{J})^{-T}\right) - \mathbf{K}_2 \quad (2.122)$$
$$Sk(V_W) = -J^T\mathbf{K}_{w_1}$$

2.3.3.7 Potential Energy

The potential energy stored in the vehicle is the sum of the contributions relative to each subsystem [55, 74, 105]. The contribution due only to the gravitational force is expressed by:

$$U_g = -\int g_0^T p_i^* \rho dv \quad (2.123)$$

The gravitational force vector is given by the difference between the lighter than air robot weight (acting at the center of gravity) and the lift force (acting upwards on the center of lift):

$$\mathbf{F}_g = \frac{\partial L}{\partial q}(q, \dot{q}) = \begin{pmatrix} (mg - B)\sin\theta \\ -(mg - B)\cos\theta\sin\phi \\ -(mg - B)\cos\theta\cos\phi \\ Bz_b\cos\theta\sin\phi - By_b\cos\theta\cos\phi \\ Bx_b\cos\theta\cos\phi + Bz_b\sin\theta \\ -By_b\sin\theta - Bx_b\cos\theta\sin\phi \end{pmatrix} \tag{2.124}$$

where B is the buoyancy force, $\bar{CN} = (x_b, y_b, z_b)^T$ represents the position of the center of lift with respect to the body fixed frame. This leaves the final kinematics/wind torque equations as

$$\mathbf{T} = \mathbf{F}(\mathbf{V}) - \mathbf{F}_g \tag{2.125}$$

Both 6 degrees of freedom models are necessary for control purpose closed loop. However, for guidance purpose, the following 3 degrees of freedom gives enough information about the lighter than air robot motion, while keeping some simplicity.

2.3.4 Translational Dynamics

For the purpose of flight path generation, it is usually sufficient to treat only the translational motion. Using a point mass model for the lighter than air robot implies that the yaw with respect to the relative wind frame is always zero. This is also demonstrated by the absence of side slip in the aerodynamic model [126].

The translational equations of an aerial vehicle through the atmosphere are directly derived from Newton's law. In this derivation, the assumptions are that the Earth is non rotating and flat, and that the vehicle weight is constant. The external forces consist of the thrust, the gravity and buoyancy forces and the aerodynamic force. The gravity and buoyancy forces are treated in the same way.

The position of the lighter than air robot is described in the local coordinate system *ENU* with unit vectors **e**, **n**, **u** pointing East, North and Up. Derivation of kinematic equations involves three velocity concepts: inertial velocity, local velocity and wind-relative velocity. The flight path coordinate system is defined, that relates the velocity vector of the vehicle with respect to Earth to the geographic system. Two angles relate the velocity coordinates to the geographic system. The heading angle χ is measured from North to the projection of V (the aircraft velocity relative to the wind) in the local tangent plane and the flight path angle γ takes vertically up to V. The wind relative velocity vector is defined by the airspeed V, the flight path angle γ and the heading χ. The variables x, y, z are the aircraft inertial coordinates. The x, y directions are chosen such that the xy plane is horizontal, the x-direction is aligned with the principal axis of symmetry and the z-direction is ascending vertically. The lighter than air equations of motion are expressed in a velocity coordinate frame attached to the lighter than air robot, considering the velocity of the wind $W = \begin{pmatrix} W_x \\ W_y \\ W_z \end{pmatrix}$ (components of the wind velocity in the inertial frame). The frame matrix being given by:

$$R = \begin{pmatrix} \cos\gamma & 0 & \sin\gamma \\ 0 & 1 & 0 \\ -\sin\gamma & 0 & \cos\gamma \end{pmatrix} \begin{pmatrix} \cos(\frac{\pi}{2}-\chi) & \sin(\frac{\pi}{2}-\chi) & 0 \\ -\sin(\frac{\pi}{2}-\chi) & \cos(\frac{\pi}{2}-\chi) & 0 \\ 0 & 0 & 1 \end{pmatrix}$$
$$= \begin{pmatrix} \cos\gamma\sin\chi & \cos\gamma\cos\chi & \sin\gamma \\ \cos\chi & -\sin\chi & 0 \\ \sin\gamma\sin\chi & \sin\gamma\cos\chi & \cos\gamma \end{pmatrix}$$

The velocity with respect to the local **ENU** frame is determined as:

$$V_{local} = V + W = R^T \begin{pmatrix} V \\ 0 \\ 0 \end{pmatrix} + \begin{pmatrix} W_x \\ W_y \\ W_z \end{pmatrix}$$

It thus may be deduced that:

$$V_{local} = (V\sin\gamma + W_z)\,\mathbf{u} + (V\cos\gamma\sin\chi + W_x)\,\mathbf{e} + (V\cos\gamma\cos\chi + W_y)\mathbf{n}$$

The translational kinematics of an aerial vehicle taking into account the wind effect can thus be expressed by the following equations:

$$\begin{aligned} \dot{x} &= V\sin\chi\cos\gamma + W_x \\ \dot{y} &= V\cos\chi\cos\gamma + W_y \\ \dot{z} &= V\sin\gamma + W_z \end{aligned} \tag{2.126}$$

Three degrees of freedom dynamical models used in trajectory optimization for aerial vehicles typically include wing body aerodynamics force effects but ignore the aerodynamic force effects produced by the control surfaces. The implementation uses the translational equations of motion formulated from Newton's law and expressed in flight path coordinates. The state variables and their derivatives are the speed of the vehicle center of mass with respect to Earth V, $\dot{V}$, the heading angle and rate χ, $\dot{\chi}$ and the flight path angle and rate γ, $\dot{\gamma}$. Newton's law is applied with aerodynamic, propulsive and gravity forces as externally applied forces.

The acceleration expressed in the local frame is [201, 231]

$$\begin{pmatrix} \dot{V} \\ 0 \\ 0 \end{pmatrix} - \begin{pmatrix} V \\ 0 \\ 0 \end{pmatrix} \times \left[\begin{pmatrix} \cos\gamma & 0 & -\sin\gamma \\ 0 & 1 & 0 \\ \sin\gamma & 0 & \cos\gamma \end{pmatrix} \begin{pmatrix} 0 \\ 0 \\ \dot{\chi} \end{pmatrix} + \begin{pmatrix} 0 \\ \dot{\gamma} \\ 0 \end{pmatrix} \right] = \begin{pmatrix} \dot{V} \\ \dot{\chi}V\cos\gamma \\ -\dot{\gamma}V \end{pmatrix}$$

The approach followed in [126, 147, 228, 231] is proposed in this section. All the external forces must be expressed in the wind axes coordinates. The contribution of the added mass phenomenon is described as:

$$F_{AM} = -\left(C_w^b\right)^T M_a C_w^b \begin{pmatrix} \dot{V} \\ \dot{\chi} V \cos\gamma \\ -\dot{\chi} V \end{pmatrix}$$

$$= -\begin{pmatrix} m_{11} & m_{12} & m_{13} \\ m_{12} & m_{22} & m_{23} \\ m_{13} & m_{23} & m_{33} \end{pmatrix} \begin{pmatrix} \dot{V} \\ \dot{\chi} V \cos\gamma \\ -\dot{\gamma} V \end{pmatrix} \tag{2.127}$$

where the added mass matrix is given by $M_a = \begin{pmatrix} X_{\dot{u}} & X_{\dot{v}} & X_{\dot{w}} \\ Y_{\dot{u}} & Y_{\dot{v}} & Y_{\dot{w}} \\ Z_{\dot{u}} & Z_{\dot{v}} & Z_{\dot{w}} \end{pmatrix}$ and the matrix C_w^b can be written as:

$$C_w^b = \begin{pmatrix} \cos\alpha & 0 & \sin\alpha \\ 0 & 1 & 0 \\ -\sin\alpha & 0 & \cos\alpha \end{pmatrix} \tag{2.128}$$

with this diagonal dominant matrix $(C_w^b)^T M_a C_w^b$ approximated by a diagonal matrix whose inverse can be calculated:

$$\begin{aligned} m_{11} &= X_{\dot{u}} \cos^2\alpha - 2X_{\dot{w}} \cos\alpha \sin\alpha + Z_{\dot{w}} \sin^2\alpha \\ m_{22} &= Y_{\dot{v}} \\ m_{33} &= X_{\dot{u}} \sin^2\alpha + 2X_{\dot{w}} \sin\alpha \cos\alpha + Z_{\dot{w}} \cos^2\alpha \end{aligned} \tag{2.129}$$

The gravity force is expressed by:

$$f_g = \begin{pmatrix} \cos\gamma \sin\chi & \cos\gamma \cos\chi & \sin\gamma \\ \cos\chi & -\sin\chi & 0 \\ \sin\gamma \sin\chi & \sin\gamma \cos\chi & \cos\gamma \end{pmatrix} \begin{pmatrix} 0 \\ 0 \\ B - mg \end{pmatrix}$$

The aerodynamic f_a and thrust f_p forces are given in body coordinates:

$$\begin{aligned} f_{ap} &= f_a + f_p \\ &= \begin{pmatrix} 1 & 0 & 0 \\ 0 & \cos\sigma & \sin\sigma \\ 0 & -\sin\sigma & \cos\sigma \end{pmatrix} \left(\begin{pmatrix} \cos\alpha & 0 & -\sin\alpha \\ 0 & 1 & 0 \\ \sin\alpha & 0 & \cos\alpha \end{pmatrix} \begin{pmatrix} T \\ 0 \\ 0 \end{pmatrix} + \begin{pmatrix} -D \\ 0 \\ L \end{pmatrix} \right) \\ &= \begin{pmatrix} T\cos\alpha - D \\ (T\sin\alpha + L)\sin\sigma \\ (T\sin\alpha + L)\cos\sigma \end{pmatrix} \end{aligned}$$

The angle σ is the bank angle.

The translational dynamics are given by the following relations:

$$\begin{aligned} \dot{V} &= \frac{1}{m + m_{11}} (T\cos\alpha - D + (B - mg)\sin\gamma) \\ &\quad - \frac{m}{m + m_{11}} \left(\dot{W}_x \cos\gamma \sin\chi + \dot{W}_y \cos\gamma \cos\chi + \dot{W}_z \sin\gamma\right) \end{aligned} \tag{2.130}$$

$$\dot{\chi} = \frac{1}{(m+m_{22})V\cos\gamma}(L + T\sin\alpha)\sin\sigma - \frac{m}{m+m_{22}}\left(\frac{\dot{W}_x\cos\chi - \dot{W}_y\sin\chi}{V\cos\gamma}\right) \tag{2.131}$$

$$\dot{\gamma} = \frac{-1}{(m+m_{33})V}(L\cos\sigma + T\cos\sigma\sin\alpha + (B-mg)\cos\gamma) + \frac{m}{(m+m_{33})V}\left(\dot{W}_x\sin\gamma\sin\chi + \dot{W}_y\sin\gamma\cos\chi + \dot{W}_z\cos\gamma\right) \tag{2.132}$$

The parameter A_r is the reference area or characteristic area, ρ is the atmospheric density, m is the mass of the lighter than air robot, M is the Mach number, $B = \rho.Vol.g$ is the buoyancy force, $C_L(M,\alpha)$ and $C_D(M,\alpha)$ are respectively the lift and drag parameters, while the angle α is the angle of attack. The lift L and drag D are given by:

$$\begin{aligned} D &= \frac{1}{2}C_D(M,\alpha)V^2A_r\rho \\ L &= \frac{1}{2}C_L(M,\alpha)V^2A_r\rho \end{aligned} \tag{2.133}$$

Atmospheric density is computed using the standard atmosphere. Generally the lift coefficient is a linear function of the angle of attack and the drag coefficient is a quadratic function of $C_L(M,\alpha)$:

$$\begin{aligned} C_L(M,\alpha) &= C_{L_0}(M) + k_{L_\alpha}(M)\alpha \\ C_D(M,\alpha) &= C_{D_0} + KC_L^2 = k_{D_0}(M) + k_{D_1}(M)\alpha + k_{D_2}(M)\alpha^2 \end{aligned} \tag{2.134}$$

where C_{L_0}, C_{L_1} and k_{D_0}, k_{D_1}, k_{D_2} are resulting coefficients with respect to α. The induced drag factor can be determined from the aerodynamic efficiency E_{max} and the zero-lift drag coefficient C_{D_0} as

$$K = \frac{1}{4C_{D_0}E_{max}^2} \tag{2.135}$$

These equations have an important place in aerospace vehicle study because they can be assembled from trimmed aerodynamic data and simple autopilot designs. Nevertheless, they give a realistic picture of the translational and rotational dynamics unless large angles and cross coupling effects dominate the simulations. Trajectory studies, performance investigations, navigation and guidance evaluations can be successfully executed with simulations of these equations.

2.4 Aerology Characteristics

The medium of flight is the relatively thin layer of air that envelopes the Earth (about 1 percent of its diameter). This atmosphere is driven into motion by Earth rotation and solar heating and is host to a variety of complex electromagnetic, chemical and thermodynamical processes.

The physical atmospheric boundary layer flows over the Earth surface, which influences the thickness and spatial growth of the layer, in such a way that a typical Atmospheric Boundary Layer extends roughly 300 m above ground level. Because the Atmospheric Boundary Layer is turbulent, a time averaged description does not account for all of the velocity fluctuations and turbulent dispersions. To model the 3D unsteady motion, a direct numerical simulation of the incompressible Navier-Stockes equation is employed for $R_e u_\tau = 270$ where R_e is the number of Reynolds based on the boundary-layer thickness and u_τ is the wall friction velocity based on the shear stress.

The motion of a lighter than air robot in turbulence is subjected to buffeting by random external forces and as a result the attitude angles and trajectory experiences random variations with time [1, 8, 108, 228, 229]. The time scale and intensity of these responses are governed by the scale and intensity of the turbulence, as well as the speed and characteristics of the vehicle. The total velocity field of the atmosphere is variable in both space and time, composed of a mean value and variations from it. The mean wind is a parameter primarily for navigation and guidance.

At landing and takeoff, lighter than air robots fly close to the ground. The presence of the ground modifies the flow past the lighter than air robot significantly so that large changes may take place in the trim and stability. The presence of the ground imposes a boundary condition which inhibits the downward flow of air normally associated with the lifting action.

Excessive surface winds can cause an unsafe situation due to three types of wind vectors: headwind, tailwind and crosswind. In addition to steady state surface winds, essentially all high winds are accompanied by gusts. Wind gusts are a quick change in the wind speed and/or direction [2].

- Windshift is a sustained change in the average wind direction of 45 degrees or more which takes place in less than 15 mn or a lesser shift if the wind speed during this period is 6 knots or greater. A wind shift could possibly turn a strong headwind into an unacceptable crosswind that suspends operations. Windshifts are typically associated with frontal passages.
- Windshear is an abrupt change in wind speed or direction, it can be vertical or horizontal. Windshear is usually associated with frontal passages, gusty winds and convective activity. Windshear becomes a hazard when an airship is close to the ground and at slow speeds, typically during the take-off and landing phases of flight.
- En route wind is not a hazard by itself. Strong headwinds will cause longer flight times which then require more fuel to be carried out and burned.
- En route turbulence can affect the maximum altitude capability of the airship. A typical flight cruising altitude is based on smooth air. Turbulence cannot be seen at any distance ahead and is only known when it is encountered. Weather radar are an excellent indication of precipitation and moisture content which when severe, can be an indication of turbulence.

- Convective induced turbulence is caused by the instability and resulting up and down drafts due to convective activity typically associated with thunderstorms. Significant Convective induced turbulence is typically associated with moderate or heavy precipitation. Airborne radar can indicate the precipitation regions where Convective induced turbulence is likely to exist. However, it exists also outside the boundaries of precipitation and both the existence and magnitude of Convective induced turbulence is impossible to predict with current technology.
- Clear air turbulence is typically caused by the windshear that exists at the boundaries between large air masses. It exists when there is no visible sign of moisture or weather event.
- Low altitude windshear is usually associated with a severe meteorological phenomenon, called the downburst. It involves a descending column of air, which then spreads horizontally in the neighborhood of the ground. This condition is hazardous because a low and slow airship might encounter a headwind coupled with a downdraft followed by a tailwind coupled with a downdraft. Ideally, what the downburst model should generate is the wind vector as a function of position in 3D field. Such a model would yield not only the wind at the origin of the body fixed frame, but also the gradients of the vertical, horizontal and lateral wind along the airship. These gradients may provide important inputs to the aerodynamic moments and hence to the attitude dynamics, during a critical portion of the landing or take-off. The domain of principal interest is a region close to the ground. The velocities involved are low and the ground boundary layer associated with the downburst can be expected to be relatively thin.
- Another dangerous form of windshear is the microburst. Microburst is an isolated wind event associated with the mature stage of a thunderstorm cell. A microburst is formed when a column of air at high altitudes quickly cools because of evaporation of ice, snow or air and becoming denser than the surrounding air, falls rapidly to the ground. On hitting the ground, this mass of air spreads radially outward in all directions. A core of such a downburst can be about 2–3 km in width and the winds generated can be as high as 150–200 km/h, the life of a microburst is typically 5–15 mn. The microburst wind may be accompanied by rain or may be dry. The wind originated at the base of a thunderstorm starts as a downburst flowing shaft of air less than 4.5 Km in diameter. As the air falls to the ground, it accelerates to speeds that can exceed 240 Km/h at its core. As the wind shift gets close to the ground, the wind slows and gets deflected radially outward from the shaft center. The outflow wind eventually dissipates. This phenomenon can be extremely dangerous especially if the airship is low and slow [115].

2.4.1 Wind Profile

The winds change considerably with altitude. The general trend is that wind speed gradually increases as the vehicle ascends through the troposphere, reaching a peak in the jet stream altitude range of 10–15 km. It then gradually decreases again to reach a minimum speed in the lower portion of the stratosphere, generally between the 18–25 km altitude. The current wind profile changes with geographic location and the time of year. Typical wind behavior can be adequately characterized based upon meteorological data, which provides a mechanism for design phase planning. The Earth's surface creates a boundary-layer effect so that winds generally increase with altitude [147, 227].

Random turbulence is a chaotic motion of the air that is described by its statistical properties. The main statistical features that need to be considered are: stationarity, homogeneity, isotropy, time and distance scalers, probability distributions, correlations and spectra.

Moreover, unlike scalar random processes in which there is only one dependent and one independent variable, turbulence is a vector process in which the velocity vector is a random function of the position vector and of time.

Recognizing that the scale of atmospheric turbulence is usually larger than the flight vehicle, in a small region, the velocity of the wind can be regarded as a linear function of distance. It then follows that in this region, the motion is a superposition of three simple basic fields: translation, rotation and strain:

$$W(r) = W_0 + \omega_A r + E_1 r + E_2 r$$

where $W(r)$ is the wind velocity at r, W_0 is its value at the origin and ω_A, E_1, E_2 are matrices containing the nine velocity gradients as elements, ω_A represents solid-body rotation of the air and E_1, E_2 are axial and shear rates of strain.

- In the case of high altitude when the scale of atmospheric turbulence is large, the assumption of uniform velocity can be made with little loss in accuracy. This is because the components of ω_A, E_1, E_2 become small as the scale becomes large.
- In small scale turbulence, close to the ground, ω_A, E_2 are important. The wind conditions that exist close to the ground (altitude < 600 m) within the Earth are very different from those at higher altitude and are governed by quite different parameters. At these low altitudes, flight path control is crucial. The principal factors that govern the wind structure (profile of mean wind, turbulence' intensities and turbulence scales) are the roughness and uniformity of the underlying terrain and the thermal stability of the atmosphere. Thermal stability is a powerful factor, in fixing turbulence characteristics at low wind speeds, but for high winds, a neutrally stable atmosphere is a good approximation

Two wind profiles are often used to model the variation in wind close to the surface:

$$W = \frac{V^*}{K}\left(\ln\left(\frac{h}{h_0}\right) - \Psi\frac{h}{L^*}\right)$$
$$W = V_r\left(\frac{h}{h_r}\right)^A \tag{2.136}$$

where W is the wind speed at altitude h, h_0 is a roughness length, K is von Karman constant, $\Psi\frac{h}{L^*}$ is a correction for non neutral stratification defined as a function of altitude and the Monin-Obukhov length L^*, the subscript r refers to some reference velocity and height and A is a power law index that varies as a function of surface friction (e.g. ≈ 0.10 for smooth surfaces such as a lake or ocean and ≈ 0.40 for an urban area with tall buildings). The reference height h_r typically varies between 10 and 100 m, depending on the location of the measurement equipment. In addition, the power law index has been found to vary as a function of the surface wind speed. Above a particular height h^*, the wind speed can be treated as constant ($h^* \geq h_r$).

Although the mean wind direction usually veers with height, the basic model assumes that the wind vector does not change direction in the boundary layer. The following cases may be studied:

- Uniform gust approximation (simplest case)
- Linear field approximation
- Correlations and spectra of gust gradients
- Unsteady aerodynamics

Discrete gusts are isolated encounters with steep gradients (horizontal or vertical) in the horizontal and vertical speed of the air. These gradients may occur at the edges of thermals and down drafts, in the wakes of structures, mountains, hills or at temperature inversions. They may also appear as rare extremes of turbulence in clouds, storms and the jet stream, possibly associated with organized structures embedded in the otherwise chaotic background.

2.4.2 Down Burst

A down burst is one mass of cold air that descends to the ground in a column. It is a particular case of down drafts. As the wind approaches the ground the wind changes its direction and radiates outward. A down burst usually picks up dust and may be visually detected. Doppler radar is able to look through a down burst and measure its air movement. A down-draft flow is classically modeled by a pair of 3D vortex rings that are placed symmetrically with respect to the ground. The stream function of a 3D irrotational and incompressible ring vortex is expressed as complete elliptic integrals. In this paragraph, the analytic formulation presented in [228] will be followed where a wind velocity components for a primary vortex ring located at (0,0, H) are modeled by:

$$\begin{aligned} W_x &= \frac{x}{r} W_m H_m \\ W_y &= \frac{y}{r} W_m H_m \\ W_z &= \frac{4}{3} W_m H_z \end{aligned} \tag{2.137}$$

with

$$H_m = \frac{h_p/r_{2p} - h_p/r_{1p}}{\sqrt{r^2 + h_p^2 + R_v^2}} - \frac{h_m/r_{2m} - h_m/r_{1m}}{\sqrt{r^2 + h_m^2 + R_v^2}} \tag{2.138}$$

$$W_m = \frac{1.182}{2\pi} R\Gamma(1 - e^{r_{1p}/r_{1m}}) \tag{2.139}$$

$$H_z = \frac{(\frac{r_1}{r_{1p}})^{0.75} - (\frac{r_2}{r_{2p}})^{0.75}}{(0.25r^2 + h_p^2 + R_v^2)^{0.75}} - \frac{(\frac{r_1}{r_{1m}})^{0.75} + (\frac{r_2}{r_{2m}})^{0.75}}{(0.25r^2 + h_m^2 + R_v^2)^{0.75}} \tag{2.140}$$

and

$$\begin{aligned} r &= \sqrt{x^2 + y^2} \qquad h_p = h + H \qquad h_m = h - H \\ r_1 &= r - R \qquad r_2 = r + R \qquad r_{1p} = r_1^2 + h_p^2 \\ r_{1m} &= r_1^2 + h_m^2 \qquad r_{2p} = r_2^2 + h_p^2 \qquad r_{2m} = r_2^2 + h_m^2 \end{aligned} \tag{2.141}$$

R_v is the ring vortex radius and Γ is its intensity. In this paragraph, the downburst wind field is assumed stationary, for example for W_x:

$$\dot{W}_x = \frac{\partial W_x}{\partial x}\dot{x} + \frac{\partial W_x}{\partial y}\dot{y} + \frac{\partial W_x}{\partial z}\dot{z} \tag{2.142}$$

The same stands for W_y, W_z.

In a surveillance flight, an aerial vehicle would circle around a given area. The onboard warning/detection systems may be: radar, lidar (light detection and ranging) or infrared type devices. Radar devices measure the Doppler velocity of water droplets moving with the winds. When there is no rain, the Lidar devices perform well because the laser signal reflects from aerosol particles carried in the atmosphere at low altitudes. The infrared devices rely on the temperature changed in the farfield (1–3 km for the lighter than air robot) caused by the windshear/microburst activity. Once a down-draft is discovered, a lighter than air vehicle may repeatedly fly through it during its limited lifespan. A downburst may last about 5–10 min. It can create vertical winds up to 20 ms/s and as it spreads out near the ground, horizontal winds up to 40 m/s. If the potential peak airspeed exceeds the performance capability of a lighter than air vehicle, it should try to avoid the down-draft. Some collision avoidance methods are presented in the next chapter.

2.5 Conclusions

Lighter than air robots are a highly interesting study object due to their stability properties. Here, motion is referenced to a system of orthogonal body axes fixed in the lighter than air robot, with the origin assumed to coincide with its bow. In the first part of this chapter, six degrees of freedom dynamic modeling of LTAR were discussed with two different approaches: Newton-Euler method and Lagrange method. The second section of this chapter addresses the problem of translational dynamics. Atmospheric dynamics, presented in the last section of this chapter, display hardly predictable behavior making the system output differ from the nominal conditions.

After establishing the equations of motion of lighter than air robots, some questions arise:

- How can be planned missions?
- How to avoid obstacles?
- How to characterize trim trajectories
- What are the handling qualities of this lighter than air robot?
- What are their controllability and stabilizability properties? How can closed loop control systems be solved?...

Answers to these questions are the topics of the following chapters.

Chapter 3
Mission Planning

Abstract Planning can be considered as the generation of a set of paths from a set of initial states to a set of goal states of a vehicle through an environment with obstacles. Many approaches have been investigated for solving these problems. All involve some kinds of simplification aiming to capture key elements of the task in a form suitable for practical computation. For many aerial robot applications, a point vehicle representation is usually used as an assumption that simplifies the problem.

3.1 Introduction

Mission planning is the process of planning how a mission will be conducted. This includes determining the set of waypoints (flight planning), the path for the lighter than air robot to fly (path planning) and the operation (task scheduling) of the payload on board the lighter than air robot. Flight planning involves creating a plan to guide a lighter than air robot from its initial position to a destination way point. A flight plan has also to consider the region where the flight will be performed and has to take a set of safety requirements into account. Mandatory safety requirements, like not running out of fuel, have to be satisfied in order to avoid the loss of the aerial robot. The ability for the lighter than air robot to manage its position away from obstacles represents a significant issue and a necessity for low altitude operations.

A vehicle is an object that is capable of motion. Generally, vehicles are represented by a position vector in 2 or 3 dimensional space and an orientation vector, along with a geometric model of the vehicle. A world space is the physical space in which a vehicle exists. A configuration is a vector of parameters that define the shape of the vehicle, most vehicles can be considered to be rigid bodies in 3D space and thus defined uniquely by 6 numbers: 3 position coordinates and 3 orientation coordinates. The set of all possible configurations of a vehicle is called the configuration space or C-space. It is often necessary to include the state of the aerial vehicle, which consists of the configuration coupled with rates of change of the configuration. For the lighter than air robot, the state is given, at least, by three position coordinates, three velocity coordinates, three orientation angles, three orientation rate angles, for a total of twelve variables. The dynamic characteristics of the aerial vehicle determine the dimension of the system, and many systems may

Y. Bestaoui Sebbane, *Lighter than Air Robots*,
Intelligent Systems, Control and Automation: Science and Engineering 58,
DOI 10.1007/978-94-007-2663-5_3,

use a reduced set of variables that adequately describe the physical state of the vehicle. It is common to consider smaller state space with coupled states, or to extend the state space to include higher order derivatives.

Planning schemes may be classified as explicit or implicit [225, 226]. An implicit method is one in which the dynamic behavior of the robot is specified then the trajectory and the actuator inputs required to go from the start configuration to the goal configuration are derived from the interaction between the robot and the environment. The best-known example of this method is the potential field method [107] and its extensions. Some other examples include the methods that applied randomized approaches [125], or graph theory [45]. These methods are presented in this chapter.

In contrast, explicit methods, presented in the next chapter, attempt to find solutions for the trajectories and actuators inputs explicitly during the motion. Explicit methods can be discrete or continuous. Discrete approaches focus primarily on the geometric constraints and the problem of finding a set of discrete configurations between the end states that are free from collisions.

This chapter consists of 8 sections. Flight planning is the subject of Sect. 3.2 while Sect. 3.3 presents a review of the most important methods of obstacle and collision avoidance. Section 3.4 provides the algorithms for planning with differential constraints. Sections 3.5 and 3.6 tackles respectively planning with uncertain winds and strong winds, both topics being under scrutiny in research currently. Task assignment is the subject of Sect. 3.7. Finally, some conclusions and perspectives are the subject of Sect. 3.8.

3.2 Flight Planning

A mission describes the operation of this vehicle in a given region, during a certain period of time while pursuing a specific objective. Way-points are locations to which the lighter than air robot is required to fly, typically given as latitude, longitude and altitude or in North, East and Up/Down coordinates. A flight plan is defined as the ordered set of waypoints executed by the lighter than air robot during a mission. It can be decomposed in phases. Each phase can be described by the coordinates of a pair of way-points and by some informations such as the speed and acceleration at which the airship is to fly between these way-points. A phase is completed when the successive way-point is reached by the lighter than air robot. Along the way, there may be a set of areas to visit and a set of areas to avoid. In addition, this mission planning strategy should be dynamic as the mission planning problem is to create a path in a dynamic environment.

Flight planning attempts to create paths that are consistent with the physical constraints of the lighter than air robot, the obstacle and collision avoidance and weighed regions. Weighed regions are regions with abnormally low or high pressure, wind speeds or any other factor affecting flight. 3D mission planning involves creating a pathfinder which helps the vehicle to reach the mission goal but also creates a path to satisfy different constraints during the mission. This pathfinder generates the path from the initial point to the mission goal and navigates the vehicle. The position, orientation and speed of the vehicle are known from the sensors and the flight management system has information about the meteorological conditions and probable obstacles to avoid. Flight planning requires an awareness of the environment in which it is operating. The assumption is made that the information required will be available.

Remark 3.1 A number of methods for manned aircrafts exists for obtaining this information. A radar based system of the airborne dependent surveillance broadcast (ADS-B) can provide the location and speed of nearby aircraft. Similarly, weather radar can also provide information for adverse weather condition. In addition to the location of the entities within the world, their dimensions must be known. Once the digital representation of the world (i.e. situational awareness) has been created, high level activities such as mission planning can be performed [115, 188, 219].

The aim is to replace the human expert with a synthetic one that can be deployed on board the aerial robot, provided with an understanding of potentially desirable trajectories qualities. It also has the ability to decide when a trajectory is close enough to be acceptable, except for small deviations, then recheck the trajectory. For example, if a trajectory satisfies all constraints except that it passes just a short distance from a known obstacle, the flight management system can reroute the path around the obstacle, interpolate the velocities needed and then check to ensure that theses changes did not create a new problem (e.g. violating a constraint on acceleration). This method combines a model of human decision-making with a computational technique for obtaining optimal or near-optimal trajectories.

Getting to a particular location is typically the aim of path planning. When flying at a constant altitude, kinematics of a lighter than air robot are analogous to those of a mobile robot. Some techniques have been implemented on 2D mobile robots operating in complex environments: Roadmap methods, cell decomposition and potential fields methods [45, 81, 97, 103, 107, 121, 141, 156, 186, 191, 194, 210]. Planning for mobile robots has been the essential step for realizing autonomous ground vehicles [65, 78, 86, 103, 107, 115, 124, 125, 168, 194]. The road map and cell decomposition methods rely on rules that are derived using the geometry of the obstacle field. Many problems such as motion planning for a number of circular or rectangular objects bounded by walls have been solved using the geometrical methods. These methods have been extended to the case of moving obstacles.

In the potential field method, an artificial potential field is assigned to the area where a robot works. The obstacles in the area are assigned a repulsive potential while the goal position of the robot is described by an attractive potential. Then the path of the robot is calculated by using the gradient of the total artificial potential. Different mathematical definitions have been used for defining the artificial potential fields and different strategies have been introduced for using the gradient of the total potential to find a path for the robot. However, these methods must all be augmented in some way to allow the use of a cost function that involves more than minimizing path length or distance from obstacles.

Autonomous aerial vehicles require fuel, power and time resources to move about their environment. To fully define a trajectory, a path (i.e. a sequence of configurations) must be augmented with velocities at least. Resources costs in the form of forces/torques and traversal times can then be computed from the governing equations of motion. To incorporate quantities such as fuel and time into a traditional path planner's cost function system state must be augmented with velocities and accelerations... An exhaustive search through a space of discretized dynamic parameter values (e.g. velocities) given constraints (e.g. limited accelerations) could theoretically be used to augment each path segment with a good or even optimal trajectory. However, computational efficiency is poor, and optimality is subject to the level of dynamic parameter discretization. A global planner, in addition to avoiding the dead-ends that may break a reactive navigation system, can take advantage of maneuvers which, although immediately very costly, may result in a lower total cost. This comes of course with a model of the world.

The simplest model of aerial robots is a point vehicle. Since in general, lighter than air robots do not have to fit into tight spaces while flying, the simplification of bounding the aerial vehicle by a rotationally-symmetrical solid has little effect on the trajectory generated by the algorithm. In many aerial autonomous vehicles problems, the differential constraints are significant since not accounting for the equations of motion may produce conservative results.

Remark 3.2 A motion planning algorithm is considered to be complete if and only if it finds a path when one exists, and returns a variable stating no path exists when none exists. It is considered to be optimal when it returns the optimal path with respect to some criterion. Any optimal planner is also complete. A sound planner is one that always guarantees the vehicle will enter the goal region and stop there without hitting any obstacle despite uncertainty in sensing and control. This definition implies that the uncertainties are bounded.

3.3 Motion Planning Algorithms Review

To this date, motion planning is still an active topic research. The canonical motion planning problem can be expressed as follows:

Consider a robot A moving in an Euclidean space $W = \Re^N$, $N = 2, 3$ called workspace. Let $O_1 \dots O_{on}$ be the fixed rigid obstacles, in W. Both the geometry of A, $O_1 \dots O_{on}$, and the positions of $O_1 \dots O_{on}$ in W are assumed to be known. Moreover, it is supposed that A is free flying that is the robot is not subject to any kinematic constraint. The motion planning problem is the following: given an initial and a final configuration of A in W, find if there exists a path, i.e. a continuous sequence of postures, that drives the lighter than air robot between the two configurations while avoiding collisions between A and the obstacles $O_1 \dots O_{on}$; report a failure if such a path does not exist.

Clearly, some of the hypotheses of this canonical problem may not be satisfied in applications. For example, the assumption that the robot is the only object in motion in the workspace rules out the relevant case of moving obstacles. Advance knowledge of obstacle geometry and placement is another strong assumption: especially in unstructured environments, typically in charge of detecting obstacles by means of its sensors, and the planning problem must therefore be solved on-line during the motion [6, 34, 67, 122, 155, 174]. The free flying robot hypothesis does not hold in nonholonomic mechanical systems, which cannot move along arbitrary paths in the workspace. All the above assumptions are introduced in order to reduce motion planning to the purely geometrical, but still quite difficult, problem of generating a collision free path.

3.3.1 Overall Problem Description

The choice of the algorithm depends on the type of problem to be solved. The most commonly-used metric is obstacle complexity, or the amount of information used and stored in a computer model of the environment. It is generally measured in terms of obstacles number, edges, or vertices. Other metrics are the fill ratio (percentage of the configuration space occupied by obstacles), along with higher order characteristics, such as mean passage width or degree of clustering of obstacles.

The purpose of this section is to present some motion planning algorithms published in the literature. Consistently keeping a safe distance from obstacles, and producing smooth paths that exhibit desirable properties (e.g. duration, energy usage) are typical requirements. Low computational complexity is therefore generally an important goal for an algorithm. A faster algorithm can allow a more rapid update of the solution [78].

3.3.2 Problem Types

There are a variety of problem types defined in the literature. A problem is considered static if knowledge of the environment is perfect and dynamic if knowledge of the environment is imperfect or changes as the task takes place. When the obstacles are fixed in space, the problem is called time-invariant and when they are allowed to move, the problem is called time-variant. The term differentially constrained (or kinodynamic [125]) means that the vehicle's equations of motion act as constraints on the path. It is possible to further categorize problems based on the assumed vehicle shape, environment type and behavior. The common problem types used in literature are described below and taken from the following references: [45, 65, 78, 97, 103, 107, 111, 118, 122, 124, 125, 168, 194, 210, 221].

3.3.2.1 Point Robot

In this problem, the vehicle is modeled as a point within the world space. Thus the configuration space is the same as the world space. Often, a vehicle is modeled by fitting it inside a bounding ball (in 2D Euclidean space this is a circle and in 3D Euclidean space a sphere), and the configuration space is simply the world space with the obstacles expanded by the radius of the vehicle's bounding ball. Thus the ball-shaped vehicle problem is the same as the point vehicle problem. This is a conservative approximation to and simplification of the mover's problem. The minimum length path is the optimal path.

3.3.2.2 Point Robot with Geometric Constraints

Planning with geometric constraints is the problem of moving an object through an obstacle field to a goal state. The vehicle is usually modeled as a rigid body, thus the configuration space has a larger dimension than the world space. A classical problem of this case is the piano mover's problem. For this kind of problem, it is usually assumed that the object has no dynamic constraints. Mover's problems measure complexity of the vehicle in addition to that of the obstacle field.

3.3.2.3 Point Vehicle with Differential Constraints

In problems with differential constraints, time and states have to satisfy the equations of motion of the vehicle. Typically the states are constrained by limits on velocity and acceleration, and sometimes also on higher-order derivatives of position, and propulsion related to flight envelope. For many aerial vehicle, this more realistic model is needed for stable control of the vehicle [22, 32, 35, 53, 59, 104, 109, 111, 136, 140, 145, 159, 178, 183–185, 189]. Optimality may be defined as minimizing the flight time between initial and goal points or energy consumed, or a mix between them.

Remark 3.3 Weighting factors are needed for cost functions that have more than one term. The shortest path that uses the least amount of fuel is often neither the shortest possible path, nor the path that uses the least fuel, but one which reaches a balance between them. The relative weights of these terms determine what sort of balance results. Substantial oversight is often required to analyze the sensitivity of solution characteristic to cost function weights and this sensitivity analysis may be specific to particular problems rather than fully generalizable.

Jogger's problem deals with the dynamic problem of a jogger with a limited field of view attempting to reach a goal position. The jogger's problem is representative of a vehicle with differential constraints, operating in a dynamic and possibly time-variant environment, with limited sensory range.

Bug problem is a special case of the jogger's problem, in the limit when the field of view goes to zero. This algorithm assumes that the robot is a point operating in a plane with a contact sensor to detect obstacles. This algorithm requires two behaviors: move on a straight line and follow a boundary. The most straightforward path planning approach is to move toward the goal, unless an obstacle is encountered in which case, circumnavigate the obstacle unless motion toward the goal is once again allowable. The robot is assumed to be a point with perfect positioning with a contact sensor that can detect an obstacle boundary if the point robot touches it. The robot can also measure the distance d(x, y) between any two points x and y. Finally, the workspace is assumed to be bounded.

3.3.2.4 General Vehicle with Differential Constraints

The differential constraints typically arise in two forms: one is on kinematics, and this kind of problem is usually called nonholonomic problem. Another one is on dynamics, involving second-order or higher differential constraints. The difference between this problem and the point robot is that now it is insufficient to model the vehicle with only a point in the world space, since six variables are needed to indicate the position of the vehicle in a three dimensional Euclidean space. For most cases, the configuration space is not a simple Euclidean space. When obstacles are involved, the configuration space itself is not adequate to represent the obstacle avoidance requirements, a higher order phase has to be employed.

In time-varying environments problem, the vehicle has to avoid obstacles that are moving in time. Optimal planners for time-varying environments generally attempt to minimize path length or time.

3.3.2.5 Obstacle Representation

Obstacles may have different shapes, convex or nonconvex. Nonconvex obstacles can be embedded into convex forms.

A 3D ellipsoid is defined by 9 parameters [223]:

- 3 coordinates of the center location x_c, y_c, z_c
- 3 semi-axes a, b, c
- 3 orientation angles ϕ, θ, ψ that relate the basic coordinate system to the principal system X, Y, Z

$$\begin{pmatrix} X \\ Y \\ Z \end{pmatrix} = R_x(\phi) R_y(\theta) R_z(\psi) \begin{pmatrix} x - x_c \\ y - y_c \\ z - z_c \end{pmatrix}$$

If a given point (x_p, y_p, z_p) after the transformation presented above to the principal system satisfies $\frac{X_p^2}{a^2} + \frac{Y_p^2}{b^2} + \frac{Z_p^2}{c^2} > 1$

It is outside of the ellipsoid obstacle.

A cuboid obstacle element is also defined by 9 parameters. A given point x_p, y_p, z_p is outside the cuboid if after the transformation, one of the following conditions is verified:

$$|X_p| > \frac{a}{2} \qquad |Y_p| > \frac{b}{2} \qquad |Z_p| > \frac{c}{2}$$

A cylinder obstacle is described by 8 elements which include coordinates of the center of the bottom surface x_b, y_b, z_b, radii of the top and the bottom surfaces r_t, r_b, the height L and 2 orientation angles ϕ, θ where the axis of the cylinder is initially aligned with the z axis. Transformation to a principal body axis is given by:

$$\begin{pmatrix} X \\ Y \\ Z \end{pmatrix} = R_x(\phi) R_y(\theta) \begin{pmatrix} x - x_c \\ y - y_c \\ z - z_c \end{pmatrix}$$

A given point (x_p, y_p, z_p) is outside of the cylinder if after the transformation, one of the following condition is met:

$$X_p^2 + Y_p^2 > \left(r_b - (r_b - r_t) \left(\frac{Z_p}{L} \right) \right)^2 \qquad H_p < 0 \qquad H_p > L$$

Superquadric functions provide an efficient and flexible means of representing geometric shapes, overcoming the deficiencies of other representations, such as spherically symmetric Gaussian or power law functions where objects are represented as spheres of diameter equal to the maximum physical object dimension.

Superquadrics are a family of complex geometric objects which include super-ellipsoids and super-hyperboloids. Superquadrics are mathematical representations of solid objects. They are a set of parametric functions that have great utility in object modeling. Their parametric characteristics enable the creation of a range of object shape by manipulating the roundness and shape parameters.

A generic superquadric function is defined in body axes as:

$$\left[\left(\frac{x_B}{a} \right)^{2/\varepsilon_2} + \left(\frac{y_B}{b} \right)^{2/\varepsilon_2} \right]^{\varepsilon_2/\varepsilon_1} + \left(\frac{z_B}{c} \right)^{2/\varepsilon_2} = 1 \tag{3.1}$$

For example, to define a spherical shape at some distance from the object edges, the shape parameters ε_1 and ε_2 should equal unity.

The inside-outside function F defines whether a point lies inside, on the surface or outside a superquadric surface and is given by:

$$F(a, X_B) = \left[\left(\frac{x_B}{a} \right)^{2/\varepsilon_2} + \left(\frac{y_B}{b} \right)^{2/\varepsilon_2} \right]^{\varepsilon_2/\varepsilon_1} + \left(\frac{z_B}{c} \right)^{2/\varepsilon_2} \tag{3.2}$$

Consider any point with coordinates (x_B, y_B, z_B) with respect to a set of body axes to the surperquadrics. If $F < 1$, point P lies inside the superquadric, whereas if $F = 1$, the point lies on the superquadric surface and finally if $F > 1$, the point lies outside the superquadric. Various obstacles shapes can now be represented using the superquadric methodology by adjusting the five parameters defined in Eq. (3.1).

3.4 Planning with Differential Constraints

Most motion planning problems relevant to autonomous aerial vehicles applications have to be considered as dynamics-constrained problems, affecting energy or duration of the trajectory. Taking into account the equations of motion is directly relevant to guaranteeing the soundness of the planner, since approximating the dynamics only through a kinematic model with constraints will lead to overly conservative models. Performance criteria have a significant effect on the resulting trajectories and heuristic methods (e.g. minimizing distance) are not able to meet specific performance requirements. This class of planning problems is more difficult to solve due to the dependency between time and the state-space introduced by the differential constraints. Even in the trivial case of connecting two states in a configuration space without obstacles, an exact solution is generally not possible.

For applications requiring a vehicle to navigate among obstacles, algorithms that exploit some form of approximation or heuristic are necessary not only for finding a feasible or sub-optimal trajectory but also for the need to negotiate with hardware capacity. Solutions to this class of problem represent a newer research area where very few approximating bounds or benchmarking results have been proposed [78].

3.4.1 Roadmap Algorithm

The road map algorithm applies sampling methods to the trajectory planning and dynamic planning problems. It handles high dimensionality and global constraints. Sampling methods are not based on a rigorous mathematical structure. Despite the existence of numerous distinct sampling techniques, they all share similar defining actions.

3.4.1.1 Cell Decomposition

The dimensions of the workspace environment are assumed to be much bigger than the dimensions of the lighter than air robot. Thus static route planning can be appropriate. The 3D environment is discretized in space over an n_x, n_y, n_z regular grid along the Cartesian directions. Let Δx, Δy, Δz be the grid intervals in the x, y, z axis respectively. Any point in the grid defines a node $n = (h_x, h_y, h_z)$, $0 \leq h_x \leq n_x$, $0 \leq h_y \leq n_y$, $0 \leq h_z \leq n_z$. A path Γ between a starting node n_s and a destination node n_d is defined through a sequence of nodes $\Gamma = n_s, \ldots, n_i, n_{i+1}, \ldots, n_d$ and it is made of straight line segments connecting any two adjacent nodes n_i, n_{i+1}. In practice, it is assumed that the navigation of the lighter than air robot is defined through via-points that are the nodes of the grid. A current velocity vector V_c is defined at any point in space. The traveling time required by a given path is evaluated computing and adding up the time required to cover each segment constituting the path. Consider the ith segment $n_{i-1}n_i$ connecting the nodes n_{i-1}, n_i of any arbitrary path, let d_i indicate its length and let $\overline{e}_i$ be a unitary vector oriented along the segment $n_{i-1}n_i$ in the direction of desired motion of the vehicle. A graph is a collection of nodes (vertices) and edges. Typically in motion planning, a node represents a salient location and an edge connects two nodes that correspond to locations that have an important relationship. This relationship could be that the nodes are mutually accessible from each other, two nodes are within line of sight of each other, two pixels are next to each other in a grid. This relationship does not have to be mutual if the robot can traverse from nodes V_1 to V_2 but not from V_2 to V_1. The edge E_{12} connecting V_1 and V_2 is directed. Such a collection of nodes and edges is called a directed graph. Typically, one searches a tree for a node with some desired properties such as the goal location for the robot. This paragraph is based mostly on references [45, 125].

There exist two main categories of search algorithms

- Uninformed methods where no information exists concerning which node should the algorithm explore next. As a result, nodes are opened one by one until a goal is reached, such as breadth first or depth first algorithms...
- Informed methods use some form of heuristics in order to select the next node to open, such as branch and bound algorithm, A* and D* algorithms...

The exact methods decompose the free configuration space into smaller convex polygons, which are then connected by a graph and searched using a graph search.

- Trapezoidal Decomposition: This approach divides the free space into trapezoidal regions by dividing it with vertical lines from each of the obstacles vertices. The vertical lines are trimmed so that they do not bisect the obstacles themselves. A road map is then formed by connecting the midpoints of adjacent trapezoids, and searched using a graph searching algorithm.
- Critical Curve Based Decomposition: While the trapezoidal decomposition is useful for point vehicle path planning, rigid vehicles with freedom to rotate require a more complex approach. In this algorithm, free space is divided into critical and non critical regions. The boundaries of these regions are piecewise polynomial curves. The various regions formed by the decomposition process are connected by a graph and this graph is searched for a path. The algorithm is for 2D problems. The Cylindrical Algebraic decomposition extends the critical-curve decomposition to 3D problems. It bisects parts of the free space using critical surfaces.
- Connected Balls in Free Space: This approach is designed to deal with unstructured obstacle fields and operates by filling free space with overlapping balls (for instance, spheres are balls in 3D Euclidean space) that are totally in free space.
- Rectanguloid Cell Decomposition: This divides the entire configuration space into rectanguloid regions, and labels each rectanguloid as being completely filled (black) partially filled (grey) or completely empty (white). The most common example is that of the A* or D* search over a square of cubic grid of occupied or unoccupied cells. This Approximate and Decompose method is similar to the trapezoidal decomposition, but replaces the triangular end regions with rectangular mixed regions. This approach reduces the proportion of mixed area in comparison with a grid decomposition with mixed cells.
- Quadtree or Octree decomposition: This decomposition is designed to reduce the number of points needed to represent obstacles as compared to a full grid representation.

In general, sampling techniques synthesize a dense tree or road map of nodes and edges. Each node represents a particular instantaneous state of the vehicle and an edge connects the two nodes via a planned path. Construction of the trees or road maps is through deterministic or random generation of new nodes and then connecting the new nodes to the old ones via a planned path that can incorporate the vehicle constraints. Once the construction phase is completed (i.e. a connection exists between start and finish), post processing on the resulting route is typically necessary and desired to ensure smoothness and improve optimality. Overall, these methods are excellent at quickly generating feasible solutions; however, complex paths and increased dimension of the system kinematic and dynamic equations can bog down the computational speed. These methods reduce the problem to that of a graph search by fitting a graph or a road map to the space.

- Visibility Graph: This is an exact solution to the 2D point vehicle problem. This approach uses the knowledge that the shortest path grazes polygonal obstacles at their vertices and builds a road map of lines connecting each vertex with all vertices visible from its position. Since the minimum-length path comes arbitrarily close to obstacles many times in a typical path, this approach offers no safety buffer to prevent collisions in the case of systems with uncertainty in their position. To avoid this problem, the obstacle space is expanded by a ball larger than the vehicle's longest radius.
- Edge-Sampled Visibility Graph: This algorithm approximately solves the 3D path length minimization point vehicle problem. This algorithm assigns multiple vertices along edges of polyhedral obstacles so that there is a minimum edge length n and builds a visibility graph from this expanded set of vertices.
- Voronoi Road map: Given the difficulty in controlling vehicles precisely enough to follow the minimum-distance path without risk of colliding with obstacles, many skeleton-based road map approaches have been taken. The Voronoi approach builds a skeleton that is maximally distant from the obstacles, and finds the minimum distance path that follows this skeleton. This algorithm is a 2D algorithm, complete but not optimal. Voronoi diagram is a special kind of decomposition of a metric space determined by distances to a specified discrete set of objects in the space. Given a set of points S, the corresponding Voronoi diagrams are generated, each point P has its own Voronoi cell which consists of all points closer to P than any other points. The border points between polygons are the collection of the points with the distance to shared generators.

3.4.1.2 Depth First and Breadth-First Algorithms

- A depth first search starts at the root, chooses a child then that node's child and so on until finding either the desired node or a leaf. If the search encounters a leaf, the search then backs up a level and then searches through an unvisited graph until the desire node is found or all nodes are visited in the graph.
- Breadth first search is the opposite. The search starts at the root then visits all of the children of the root first. Next, the search then visits all of the grand children and so forth. The belief here is that the target node is near the root so this search

would require less time. In 2D, Four point connectivity will only have edges to the North, South, East and West, whereas 8 point connectivity will have edges to all pixels surrounding the current pixel: North, East, South, West, North-East, North-West, South-East, South-West. In 3D, there are 26 neighbors for a 45 degrees discrimination [54].
- The wave-front planner is an implementation of a breadth first search. In general, a breadth first search is implemented with a list in a First-In First-Out (FIFO) manner, a queue. The depth first search contrasts in that the nodes are placed in a Last-In First-Out (LIFO) manner, a stack. Another common search is called a greedy search which expands nodes that are closest to the goal. Here the data structure is called a priority queue in that the nodes are placed into a sorted list based on a priority value. This priority value is a heuristic that measures distance to the goal node.

Remark 3.4 Breadth-first search produces the shortest path to the start node in terms of link strengths. Since the wavefront planner is a breadth-first search, a four point connectivity wave front algorithm produces the shortest path with respect to the distance function. It has an underlying graph, where each node corresponds to a pixel and neighboring pixels have an edge length of one.

The graph search can be tuned to find optimal paths with respect to metrics such as energy, time, distance traversability, safety... as well as combinations of them. There is also the efficiency: minimize the number of nodes that have to be visited to locate the goal node subject to the path optimality criteria. Depth first and breadth first are uninformed: the search just moves through the graph without any preference for or influence on where the goal node is located. For example, if the coordinates of the goal node are known, then a graph search can use this information to help decide which nodes in the graph to visit (i.e. expand) to locate the goal node. A graph search may choose as its next node to explore one that has the shortest Euclidean distance to the goal because such node has highest possibility, based on local information, of getting closest to the goal. However, there is no guarantee that this node will lead to the globally shortest path in the graph to the goal. This is just a good guess. However, these good guesses are based on the best information available to the search.

3.4.1.3 A* Algorithm

The A* algorithm searches a graph efficiently with respect to a chosen heuristic. The A* algorithm will return an optimal path is the heuristic is optimistic. An optimistic or admissible heuristic always returns a value less than or equal to the cost of the shortest path from the current node to the goal node within the graph. The input for A* is the graph itself. These nodes can naturally be embedded into the lighter than air robot free space and thus have values corresponding to the cost required to traverse between the adjacent nodes. The output of the A* algorithm is a back pointer path, which is a sequence of nodes starting from the goal and going back to the start.

The A* algorithm has a priority queue which contains a list of nodes sorted by priority, determined by the sum of the distance traveled in the graph thus far from the start node and the heuristic. The first node to be put into the priority queue is naturally the start node. Next, the start node is expanded by popping the start node and putting all adjacent nodes to the start node into the priority queue sorted by their corresponding priorities.

The input for the A* algorithm being the graph itself, the nodes can naturally be embedded into the robot's free space and thus can have coordinates. Edges correspond to adjacent nodes and have values corresponding to the cost required to traverse between the adjacent nodes. The output of the A* algorithm is a back pointer path, which is a sequence of nodes starting from the goal and going back to the start. Two additional structure are used, an open set O and a closed set C. The open set O is the priority queue and the closed set C contains all processed nodes. Other notation includes

- $Star(n)$ represents the set of nodes which are adjacent to n
- $C(n_1, n_2)$ is the length of edge connecting n_1 and n_2
- g(n) is the total length of a back pointer path from n to q_{goal}
- h(n) is the heuristic cost function which returns the estimated cost of shortest path from n to q_{goal}
- $\mathrm{f(n)} = \mathrm{g(n)} + \mathrm{h(n)}$ is the estimated cost of shortest path from q_{start} to q_{goal} via n

The Euclidean distance between the current point and the destination goal, divided by the maximum possible nominal speed can be employed as a heuristic function. This choice ensures that the heuristic cost will always be lower than the actual cost to reach the goal from a given node and thus the optimum solution is guaranteed [22].

The algorithm can be formulated as follow [45]:

Algorithm A*

- Input: A graph
- Output: A path between start and goal nodes
- Repeat
    1. Pick n_{best} from 0 such that $f(n_{best}) < f(n)$
    2. Remove n_{best} from O and add to C
    3. If $n_{best} = q_{goal}$, EXIT
    4. expand n_{best}: for all $x \in Star(n_{best})$ that are not in C
    5. if $x \notin O$ then
    6. add x to O
    7. else if $g(n_{best}) + C(n_{best}, x) < g(x)$ then
    8. update x's back pointer to point to n_{best}
    9. end if
- Until O is empty

There are variations or special cases of A* algorithm.

- When $f(n) = h(n)$, then the search becomes a greedy search because the search is only considering what it believes is the best path to the goal from the current node.
- When $f(n) = g(n)$ the planner is not using any heuristic information but rather growing a path that is shortest from the start until it encounters the goal. This classic search is called Dijkstra algorithm.

3.4.1.4 D* Algorithm

In dynamic environments, there are three types of dynamic obstacles.

- ones that move significantly slower than the robot
- those that move at the same speed
- obstacles that move much faster than the robot

The super-fast obstacle case is easy to ignore because the obstacles will be moving so fast, that there probably is no need to plan for them because they will either move too fast for the planner or they will be in and out of the robot's path so quickly that it does not require any consideration. In this paragraph, dynamic environments where the world change at a speed much slower than the robot, are considered.

The A* algorithm can be run to determine a path from start to goal and then follow that path until an unexpected change occurs. The D* algorithm is devised to locally repair the graph allowing efficient updated searching in dynamic environments, hence the term D*. D* initially determines a path starting with the goal and working back to the start using a slightly modified Dijkstra's search.

The modification involves updating a heuristic and a minimum heuristic function. Each cell contains a heuristic cost (h) which for D* is an estimate of path length from the particular cell to the goal, not necessarily the shortest path length to the goal as it was for A*. These h values will be updated during the initial Dijkstra search to reflect the existence of obstacles. The minimum heuristic values h are the estimate of the shortest path length to the goal. Both the h and the values will vary as the D* search runs, but they are equal upon initialization [45].

Notation

- X represents a state
- O is the priority queue
- L is the list of all states
- S is the start state
- t(x) is the value of state with respect to priority queue

  1. t(x): New if x has never been in O
  2. t(x): Open if x is currently in O
  3. t(x): Closed if x was in O but currently is not

Algorithm D*

- **Input**: List of all states L
- **Output**: The goal state, if it is reachable, and the list of states L are updated so that back pointer list describes a path from the start to the goal. If the goal state is not reachable, return **NULL**
  - For each $X \in L$ do
  - $t(X) = New$
  - *endfor*
  - $h(G) = 0; 0 = \{G\}; X_c = S$
  - The following loop is Dijkstra's search for an initial path.

    1. repeat
    2. $k_{\min} =$ process-state(0, L)
    3. until ($k_{\min} > h(x_c)$) or ($k_{\min} = -1$)
    4. P = Get-Pointer-list(L, X_c, G)
    5. If P = Null then
    6. return (Null)
    7. end if
    8. end repeat

  - *endfor*
  - X_c is the second element of P Move to the next state in P
  - P = Get-Back-Pointer-List(L, X_c, G)
  - until X_c = G
  - return (X_c)

  Subroutine: Get-back-pointer-list(L, S, G)
  - **Input**: A list of states L and two states (start and goal)
  - **Output**: A list of states from start to goal as described by the back pointers in the list of states L

    1. If path exists then
    2. return (the list of states)
    3. else
    4. return NULL
    5. endif

  Subroutine: Insert (O, X, h_{new})
  - **Input**: Open list, a state and a h-value
  - **Output**: Open list is modified

    1. If t(X) = New then
    2. $h(X) = h_{new}$
    3. else if t(X) = Open then
    4. k(X) = min(h(X), h_{new})
    5. endif
    6. h(X) = h_{new}
    7. t(X) = Open

8. Sort O based on increasing k values

C(X,Y) is the estimated path length between adjacent states X, Y. h(X) is the estimated cost of a path from X to Goal (heuristic). k(X) is the estimated cost of a shortest path from X to Goal (minimum heuristic = $\min h(X)$ before X is put on O, values h(X) takes after X is put on O). $b(X) = Y$ is the measured distance adjacent states with X, Y.

Subroutine

- **Input**: The Open list, 2 states and a value
- **Output**: A k value and the open list gets updated

1. $C(X,Y) = c_{val}$
2. If t(X) = Closed then
3. Insert (O, X, h(X)) = min(h(X), h_{new})
4. endif
5. h(X) = h_{new}
6. t(X) = Open
7. Sort O based on increasing k values

Subroutine: Min – State(0)

- **Input**: The Open list O
- **Output**: The state with minimum k value in the list related values

1. if $O = \emptyset$ then
2. Return (−1)
3. else
4. return ($\text{argmin}_{Y \in O} k(Y)$)
5. endif

Subroutine: Get – Min(0)

- **Input**: The Open list O
- **Output**: Lowest k-values of all states in the open list

1. if $O = \emptyset$ then
2. Return (−1)
3. else
4. return ($\text{argmin}_{Y \in O} k(Y)$)
5. endif

Subroutine: Process – State

- **Input**: List of all states L and the list of all states that are open
- **Output**: $k_{\min}$ and an updated list of all states and an updated open list

1. If X = Min – State(O)
2. If X = Null then
3. return (−1)
4. endif
5. $k_{old} = Get - Min(O)$
6. delete (X)

7. $k_{old} < h(X)$ then
8. for each neighbor $Y \in L$ of X do
9. If $h(Y) \leq k_{old}$ and $h(X) > h(Y) + C(X, Y)$ then
10. b(X) = Y
11. h(X) = h(Y) + C(X, Y)
12. endif
13. endfor
14. endif
15. if K_{old} = h(X) then
16. for each neighbor $Y \in L$ of X do
17. if (t(Y) = New) or ($b(Y) = X$ and $h(Y) \neq h(Y) + C(X, Y)$) or ($b(Y) \neq X$ and $h(Y) > h(Y) + C(X, Y)$) then
18. b(Y) = X
19. Insert $(O, Y, h(X) + C(X, Y))$
20. endif
21. endfor
22. else
23. for each neighbor $Y \in L$ of X do
24. If (t(Y) = New) or (b(Y) = X and h(Y) $\neq$ h(X) + C(X, Y))
25. then b(Y) = X
26. Insert $(O, Y, h(XW) + C(X, Y))$
27. else if ($b(Y) \neq X$ and $h(X) > h(Y) + C(X, Y)$ and $t(Y) = Close$ and $h(Y) > k_{old}$ then
28. Insert $(O, Y, h(Y))$
29. endif
30. endfor
31. endif
32. Return Get-kmin (O)

The next section presents another well known method for obstacle avoidance: Artificial Potential methods.

3.4.2 Artificial Potential Methods

In many applications such as surveillance and monitoring, the lighter than air robot must be able to plan its motion on-line, i.e. using partial information on the workspace gathered during the motion on the basis of sensor measurements. An effective method for on-line planning relies on the use of artificial potential fields. Essentially, the point that represents the robot in configuration space moves under the influence of a potential field U obtained as the superposition of an attractive potential to the goal and a repulsive potential from the C obstacle region. Planning takes place in an incremental fashion: at each robot configuration Q, the artificial force generated by the potential is defined as the negative gradient $-\nabla U(Q)$ of the potential, which indicates the most promising direction of local motion. The potential field can be used to serve as a controller. The same properties apply here: the function is generally simple to compute, but may result in an incomplete planner and is non-optimal in general. If the potential field is designed properly, it may be used directly as part of a feedback controller. However, if used in that way, care needs to be taken so that the feedback controller is stable. Furthermore, to use such a method, there usually exist certain constraints on the form of the vehicle dynamics. The free configuration space can be decomposed into convex cells and then local control policies are designed for each cell to respect dynamic constraints. The convergence can be proved for a double integrator. It can also be applied for an affine system within a polyhedral environment. Such configuration space division techniques also enable a combination of control method and powerful logic-based Artificial Intelligence methods.

Potential field methods are based on the idea of assigning a potential function to the free space and simulating the vehicle as a particle reacting to forces due to the potential field. The goal point has the lowest potential, and attracts the vehicle while obstacles repel the vehicle. Since their initial publication [107], potential field methods have been generally known for being of low computational complexity but incomplete. However, a potential field which has the properties of a navigation function makes a complete path planner. The value of a potential function can be viewed as energy and hence the gradient of the potential is a force, which points in the direction that locally maximally increases the potential. The combination of repulsive and attractive forces should direct the robot from the start location to the goal location while avoiding obstacles. The robot terminates motion when it reaches a point where the gradient vanishes. Such a point is called a critical point. When the Hessian is non singular, the critical point is non degenerate, isolated. Although the methodology is appealing due to its intuitive nature and computationally efficient implementation (controls are typically available in analytic form) there is often no guarantee that local minima are not present which may trap the vehicle in some configuration other than the desired one. This problem can be overcome by generating the potential field as a numerical solution to the Laplace equation or by various heuristics such as adding noise to escape from any local minima.

There are two classes of potential fields known to satisfy properties of a navigation function: those based on a harmonic function and those based on solving the optimal distance-to-go function. These methods require, however, discretizing the configuration space into a grid with M points. The added advantage of a navigation function is that it can be used to provide direct feedback control, rather than relying on feed-forward control, as traditional trajectory planners do. A single navigation function produces a trajectory for every possible starting point in the configuration space. Both methods require full knowledge of the configuration space prior to the planning event.

- Potential Field with Gradient Descent (Virtual Force Field VFF): This original potential field approach is designed to run quickly. It assigns a decaying function to the goal point with a negative minimum value and a decaying function to each of the obstacles with a positive maximum value and sums the functions from the goal and all obstacles to get the total potential. The Virtual Force Field algorithm is sometimes used directly for trajectory generation and is valid for an arbitrary number of dimensions, the limitations are trap situations due to local minima and no passage between closely spaced obstacles.

- Harmonic Potential Functions: This class of functions is based on solving a partial differential equation with a Laplacian term. These equations include Laplace's equation, Poisson's equation, the conduction heat flow equation,... While not producing an optimal path, these equations generate functions that are smooth navigation functions, have only one local minimum that occurs at the goal point, potential obtaining a constant maximal value at the boundaries of obstacles, and has a non-degenerate Hessian at each critical point of the function.

Remark 3.5 The method of Potential Field Guided Search (Depth-First, Best-First, Variational Planning-Arbitrary Potential Field) assumes limited computational requirements and the need to react under changing environments. An intuitive visualization of this method is to imagine the vehicle to be a positively charged particle in the state space. Obstacles have a similar positive charge—repulsive force—and the target location has a negative charge—attractive force. The accumulation of these forces drives the vehicle towards the goal. This approach is designed for potential fields that have local minima. Rather than use gradient descent, which is easily trapped in local minima, a search that is complete in the resolution or probabilistic sense is used. This can be considered as being similar to an A* search with the simple heuristic replaced by a potential field. The variational planning approach uses the potential as a cost functional and attempts to find a path to the goal point that minimizes this cost.

3.4.2.1 Classical Methods

As noted before, the artificial potential field methodology is based on the assumption of the existence of a virtual potential field which attracts the vehicle towards a goal, while repelling it away from obstacles and other flying vehicles. This section is based mainly on [65, 109, 194].

Attractive Potential The attractive potential is designed so as to guide the robot to the goal configuration Q_g. To this end, a paraboloid function may be used:

$$U_{a1}(Q) = \frac{1}{2} k_a \left\| e(Q) \right\|^2 \tag{3.3}$$

where $k_a > 0$ and $e = Q_g - Q$ is the error vector with respect to the goal configuration Q_g. This function is always positive and has a global minimum in Q_g, where it is zero. The resulting attractive force is defined as:

$$f_{a1}(Q) = -\nabla U_{a1}(Q) = -k_a e(Q) \tag{3.4}$$

Hence, f_{a1} converges linearly to zero when the robot configuration q tends to the goal configuration Q_g.

Alternatively, a conical attractive potential may be defined as:

$$U_{a2}(Q) = k_a \|e(Q)\| \tag{3.5}$$

Also, U_{a2} is always positive and zero in Q_g. The corresponding attractive force is

$$f_{a2}(Q) = -\nabla U_{a2}(Q) = -k_a \frac{e(Q)}{\|e(Q)\|} \tag{3.6}$$

that is constant in modulus. This represents an advantage with respect to the force f_{a1} generated by the paraboloid attractive potential, which tends to grow indefinitely as the error vector increases in norm. On the other hand, f_{a2} is indefinite in Q_g. A choice that combines the advantages of the above two potentials is to define the attractive potential as a conical surface away from the goal and as a paraboloid in the vicinity where $\|e(Q)\| = 1$ (i.e. on the surface of the sphere of unit radius centered in Q_g), one obtains an attractive force that is continuous for any Q.

Repulsive Potential The repulsive potential U_r is added to the attractive potential U_a to prevent the lighter than air robot from colliding with obstacles as it moves under the influence of the attractive force f_a. In particular, the idea is to build a barrier potential in the vicinity of the C obstacle region, so as to repel the point that represents the robot. In the following, the C obstacle region has been partitioned in convex components CO_i, $i = 1 \dots p$. These components may coincide with the C obstacles themselves. In the presence of non convex C obstacles, it is necessary to perform the decomposition in convex components before building the repulsive potential.

For each convex component CO_i, an associated repulsive potential is defined as:

$$U_{r,i} = \begin{cases} 0 & \text{if } \eta_i(Q) > \eta_{0,i} \\ \frac{k_{r,i}}{2}\left(\frac{1}{\eta_i(Q)} - \frac{1}{\eta_{0,i}}\right)^2 & \text{if } \eta_i(Q) \le \eta_{0,i} \end{cases} \tag{3.7}$$

where $\eta_{0,i}$ is the range of influence of CO_i, $k_{r,i} > 0$, $\eta_i(q)$ is the distance of q from CO_i

$$\eta_i(Q) = \min_{Q' \in CO_i} \|Q - Q'\| \tag{3.8}$$

The potential $U_{r,i}$ is zero outside CO_i and positive inside the range of influence $\eta_{0,i}$ and tends to infinity as the boundary of CO_i is approached.

- When $C = \Re^2$ and the convex component CO_i is polygonal, an equipotential contour of $U_{r,i}$ (i.e. the locus of configurations q such that $U_{r,i}$ has a certain constant value) consists of rectilinear tracts that are parallel to the sides of the polygon, connected by arcs of circle in correspondence of the vertices. The contours get closer to each other in the proximity of the C obstacle boundary, due to the hyperboloidic profile of the potential.
- When $C = \Re^3$ and the convex component CO_i is polyhedral, the equipotential surfaces of $U_{r,i}$ are copies of the faces of CO_i, connected by patches of cylindrical surfaces in correspondence of the edges and spherical surfaces in correspondence of the vertices of CO_i.

The repulsive force resulting from $U_{r,i}$ is:

$$f_{r,i} = -\nabla U_{r,i}(Q) = \begin{cases} 0 & \text{if } \eta_i(Q) > \eta_{0,i} \\ \frac{k_{r,i}}{\eta_i^2(Q)}\left(\frac{1}{\eta_i(Q)} - \frac{1}{\eta_{0,i}}\right)\nabla\eta_i(Q) & \text{if } \eta_i(Q) \leq \eta_{0,i} \end{cases} \tag{3.9}$$

Denote by Q_m the configuration of CO_i that is closer to Q (Q_m is uniquely determined in view of the convexity of CO_i). The gradient vector $\nabla\eta_i(Q)$ which is orthogonal to the equi-potential contour (or surface) passing through q. If the boundary of CO_i is piecewise differentiable, the function η_i is differentiable everywhere in C_{free} and $f_{r,i}$ is continuous in the same space.

The aggregate repulsive potential is obtained for the *on* obstacles by:

$$U_r(Q) = \sum_{i=1}^{on} U_{r,i}(Q) \tag{3.10}$$

If $\eta_i(Q_g) > \eta_{0,i}$ for $i = 1..on$ (i.e. if the goal is placed outside the range of influence of each obstacle component CO_i), the value of the aggregate repulsive field U_r is zero in Q_g. In the following, it will be assumed that this is the case.

Total Potential The total potential U_r is obtained by adding the attractive and the aggregate repulsive potentials:

$$U_t(Q) = U_a(Q) + U_r(Q) \tag{3.11}$$

This results in the force field

$$f_t(Q) = -\nabla U_t(Q) = f_a(Q) + \sum_{i=1}^{on} f_{r,i}(Q) \tag{3.12}$$

$U_t(Q)$ clearly has a global minimum in q_g, but there may also exist some local minima where the force field is zero. This happens in the 'shadow zone' when the repulsive potential $U_{r,i}$ has equi-potential contours with lower curvature than the attractive potential in the same area

Planning Techniques Different approaches for planning collision-free motions are briefly discussed below [194]:

1. The first possibility is to let

$$\tau = f_t \tag{3.13}$$

hence considering $f_t(q)$ as a vector of generalized forces that induce a motion of the robot in accordance with its dynamic model.

2. The second method regards the robot as a unit point mass moving under the influence of $f_t(q)$ as in

$$\ddot{Q} = f_t \tag{3.14}$$

3. The third possibility is to interpret the force field $f_t(q)$ as a desired velocity for the robot by letting

$$\dot{Q} = f_t \tag{3.15}$$

In principle, one could use these three approaches for on-line as well as off-line motion planning. In the first case, Eq. (3.13) directly represents control inputs for the robot, whereas the implementation of Eq. (3.14) requires the resolution of the inverse dynamics problem. Equation (3.15) can instead be used on-line in a kinematic control scheme, in particular to provide the reference inputs for the low-level controllers that are in charge of reproducing such generalized velocities as accurately as possible. In any case, the artificial force field f_t represents, either directly or indirectly, a true feedback control that guides the robot towards the goal, that have been detected by the sensory system. To emphasize this aspect, on-line motion generation based on artificial potentials is also referred to as reactive planning.

In off-line motion planning, configuration space paths are generated by simulation. In general, the use of Eq. (3.13), generates smoother paths, because with this scheme, the reactions to the presence of obstacles are naturally filtered through the robot dynamics. On the other hand, the strategy represented by Eq. (3.15) is faster in executing the motion corrections suggested by the force field f_t and thus be considered safer. The characteristics in Eq. (3.14) are intermediate between the other two. Another aspect to be considered is that using Eq. (3.15) guarantees (in the absence of local minima) the asymptotic stability of Q_g (i.e. the robot reaches the goal with zero velocity) whereas this is not true for the other two motion generation strategies. To achieve asymptotic stability with Eqs. (3.13), (3.14), a damping term proportional to the robot velocity $\dot{Q}$ must be added to f_t. The most common choice is the simple numerical integration of Eq. (3.15) via the Euler method:

$$Q_{k+1} = Q_k + T f_t(Q) \tag{3.16}$$

where Q_{k+1}, Q_k represent respectively the current and the next robot configuration, and T is the integration step. To improve the quality of the generated path, it is also possible to use a temporal variable T, smaller when the modulus of the force field f_t is larger in the vicinity of obstacles) or smaller (close to the destination Q_g). Equation (3.16) may be interpreted as a numerical implementation of the gradient method for the minimization of $U_t(Q)$, often referred to as the algorithm of steepest descent.

Local Minima Problem The possibility of the existence of local minima in the artificial potential field could be one of the drawbacks of the potential field method. A local minimum can attract and trap the robot, preventing it from reaching its final goal. Search methods have been introduced to address this problem at a high com putational cost. One method for avoiding the generation of local minima is adding multiple auxiliary attraction potentials, whose positions are determined by a genetic algorithm. Also a set of analytical guidelines have been given for designing potential functions to avoid local minima for a number of representative scenarios. Another approach is based on the use of navigation functions, i.e. artificial potentials that have no local minima.

Definition 3.1 A function $\Phi : Q_{free} \rightarrow [0, 1]$ is called a navigation function if it

- is smooth (or at least C^k for $k \geq 2$)

- has a unique minimum at Q_g in the connected component of the free space that contains Q_g
- is uniformly maximal on the boundary of the free space
- is Morse (A Morse function is one whose critical points are all non degenerate, critical points are isolated.)

A way to define a navigation function consists of building first a diffeomorphism that maps the C obstacle region to a collection of spheres, then generating a classical total potential in the transformed space, and finally mapping it back to the original configuration space so as to obtain a potential free of local minima. If the C obstacles are star-shaped, such a diffeomorphism actually exists, and the procedure outlined above provides in fact a navigation function. Another possibility is to build the potential using harmonic functions, that are the solutions of a particular differential approach that describes the physical process of heat transmission or fluid dynamics.It has been shown that harmonic potential do not suffer from local minima and lead to unique solutions. This property of harmonic potential functions allows the potential to be defined in Euclidean space rather than in the configuration space.

3.4.2.2 Harmonic Functions

Harmonic potential fields have been utilized for mobile robots in known environments containing stationary or moving obstacles by employing the panel method known in fluid mechanics. The panel method is currently extended to the case of unknown environments and 3D environments.

Potential theory is used in describing many conservative systems in an irrotational fluid flow. In the absence of viscous effects and rotational force, the originally irrotational flow will remain so in the region around a body inside the flow field. Let us denote the vectorial velocity field in this region by V. Vorticity vanishes when the flow is irrotational, that is:

$$V = -\nabla\Phi = \frac{\partial\Phi}{\partial x_1}\mathbf{i} + \frac{\partial\Phi}{\partial x_2}\mathbf{j} + \frac{\partial\Phi}{\partial x_3}\mathbf{k} \tag{3.17}$$

where Φ is a scalar velocity potential. Furthermore, when the fluid is incompressible, the velocity field must satisfy the continuity equation

$$V.V = 0 \Rightarrow \nabla^2\Phi = 0 \tag{3.18}$$

∇^2 is the Laplacian operator. This equation is called the Laplace or potential equation and its solutions are called harmonic or potential functions. Many physical problems are described by the Laplace equation. The properties of harmonic functions related to local minimum are described in the following.

- The Superposition property of the potential functions is related to the linearity of the Laplace equation. If Φ_1, Φ_2 are harmonic functions, then any linear combination of Φ_1 and Φ_2 is also a harmonic function and a solution of Laplace equation.

- The Mean Value Property of A 2D potential function $\Phi(x_1, x_2)$ that is harmonic in a circle with center at (x_{0_1}, x_{0_2}) is such that

$$\Phi\left(x_{0_1}, x_{0_2}\right) = \frac{1}{2\pi}\int_0^{2\pi}\left(x_{0_1} + r\cos\theta, x_{0_2} + r\sin\theta\right) d\theta \tag{3.19}$$

  The value of the potential at the center of any arbitrary circle is equal to the average of the potential integrated over the circumference of the circle. This property is independent of the radius r of the circle only if the function is harmonic inside the circle. A similar results holds for an arbitrary number of dimensions.
- The minimum/maximum property of a non constant harmonic function occurs on singular boundaries where the potential tends to infinity. The above properties of a harmonic function are very useful in building an artificial potential field for obstacle avoidance problem, eliminating the possibility of generating a stationary point in the velocity field except the goal point.

3.4.2.3 3D Harmonic Potential Functions

Uniform Label A uniform flow from a starting position to a goal position is used to set up the potential field to generate a more effective force. When the start point is far from the goal point, the goal point's potential is too weak to attract the robot effectively. The uniform flow drives the lighter than air robot toward the goal point in such a situation, resulting to a more effective potential. The 3D potential of uniform flow Φ_u is written as [21, 65]:

$$\Phi_u = -\left(a_1x_1 + a_2x_2 + a_3x_3\right)U \tag{3.20}$$

where a_1, a_2, a_3 are the direction cosines of a line connecting the start point to the goal point and U is the strength of the potential. The potential function defined for Φ_u satisfying the 3D form of the Laplace equation, is harmonic.

Goal Sink Since the robot must reach the goal, an attractive harmonic potential is needed at the goal, where the superposed potential has only one global minimum. This attractive goal can be represented by a singular point sink. The 3D harmonic potential generated by the goal sink is expressed as

$$\Phi_g = -\frac{\lambda_g}{R_g} \tag{3.21}$$

where R_g is the distance between the point (x_1, x_2, x_3) (the current location of the robot) and the goal point $G(x_{g_1}, x_{g_2}, x_{g_3})$ and λ_g is the goal sink strength. The goal potential Φ_g given by (3.21) satisfies the Laplace equation (3.18).

Spatial Panel Any 3D obstacle can be approximated by a number of spatial panels. The potential due to the obstacles can be obtained by calculating the potential for each polygonal panel and superposing the results. Harmonic sources or sinks of uniform density, similar to the functions defined for the goal potential, are distributed on each panel. A single panel is a planar on which uniform sources or sinks are distributed. The total potential resulting from this distribution must be calculated for each 3D panel. The calculation is done by integration.

The unit vector $\hat{\mathbf{n}}$ is perpendicular to the surface of the panel and is directed outward of the obstacle volume. The unit vector $\hat{\mathbf{i}}$ is directed along a leg of the panel generally shown as ∂S. The unit vector $\hat{\mathbf{u}}$ is in the plane of the panel and is perpendicular to the unit vector $\hat{\mathbf{i}}$. Panel j, with surface S_j produces the following potential at a point $C(x_1, x_2, x_3)$

$$\Phi_j(x_1, x_2, x_3) = \lambda_j \int_{S_j} \frac{dS_j}{R_j} \tag{3.22}$$

where R_j is the distance of the point $C(x_1, x_2, x_3)$ to an arbitrary point on the spatial panel j. The origin of the coordinate system is located at point O. The vector representing point C, at which the potential is being calculated is denoted by r_c. If r′ is the vector pointing from the origin at an arbitrary point on leg ∂S_k of the polygon S, then R_k can be expressed as:

$$\frac{1}{R_k} = \frac{1}{|r_c - r'|} \tag{3.23}$$

The integral (3.22) can be converted to a summation of integrals over the boundary of the panel S in which the position of an arbitrary point on the edges of the panel, Eq. (3.23) appear:

$$\Phi_j = \lambda_j \sum_k \hat{P}_k^0 \hat{u}_k \left(P_k^0 \ln \frac{R_+^k + l_+^k}{R_-^k + l_-^k} \right) - |d|\, T_1 \tag{3.24}$$

where

$$T_1 = atan \frac{P_k^0 l_+^k}{(R_k^0)^2 + |d|\, R_+^k} - atan \frac{P_k^0 l_-^k}{(R_k^0)^2 + |d|\, R_-^k} \tag{3.25}$$

More informations on this derivation can be found in [220].

All these geometric parameters can be evaluated when the edges of the panel S_i are defined. An edge k of the panel S_i is defined as a line segment by specifying the position of the two ends of the edge k with respect to the origin O as vectors r_+^k and r_-^k respectively. A unit vector $\hat{u}_k$ is defined such that it is perpendicular to the edge k and lies in the plane of the polygon S_i.

With the position of the two ends of the edge k given the distance R_+^k and R_-^k can be found:

$$R_k^0 = \frac{|(r_k^+ - r_k^-) \times (r_c - r_k^-)|}{|(r_k^+ - r_k^-)|} \tag{3.26}$$

where

$$R_k^+ = \left| r_k^+ - r_c \right| \tag{3.27}$$

$$R_k^- = \left| r_k^- - r_c \right| \tag{3.28}$$

The distance d, height of the observation point C above the plane of S_k and measured positively in the direction of $\hat{n}$, may be calculated as:

$$d = n.\left(r - r_k^+\right) = n.\left(r - r_k^-\right) \tag{3.29}$$

The three distances R_k^0, d and P_k^0 form a right angled triangle, thus:

$$P_k^0 = \sqrt{(R_k^0)^2 - d^2} \tag{3.30}$$

The signed distances l_k^- and l_k^+ from point D on the extension of the edge k to the two ends of the edge k must be calculated. The sign of these distances is positive if the vector pointing from D to the corresponding end of the line segment has the same direction $\hat{i}_k$. The signed distances l_k^- and l_k^+ can be determined after the position of point D has been calculated.

The position of point D is obtained by following the procedure discussed below. First, the unit vector of the edge k is defined as:

$$\hat{I}_k = \frac{r_k^+ - r_k^-}{|r_k^+ - r_k^-|} \tag{3.31}$$

Then the position of point D can be found as:

$$r_{D_k} = \left[\hat{I}_k (r_c - r_k^-)\right] \hat{I}_k + r_k^- \tag{3.32}$$

using the position of point D from Eq. (3.32), the following relation for l_k^- and l_k^+ can be found:

$$l_k^- = (r_k^- - r_{Dk})\hat{I}_k \tag{3.33}$$

$$l_k^+ = (r_k^+ - r_{Dk})\hat{I}_k \tag{3.34}$$

Finally, the unit vector P_k^0 can be written as

$$\hat{P}_k^0 = \frac{(r_{Dk} - r_c)\hat{u}_k}{|(r_{Dk} - r_c).\hat{u}_k|}\hat{u}_k \tag{3.35}$$

where

$$\hat{u}_k = \hat{n}_k \times \hat{I}_k \tag{3.36}$$

At the configuration proposed for the position of point C, both vectors P_k^0 and $\ddot{u}_k$ have the same direction. If the projection of point C on the panel S_j was on the other side of the edge k, both vectors P_k^0 and $\hat{u}_k$ would have opposite directions.

3.4.2.4 3D Robust Harmonic Potential Field

A 3D potential field for obstacle avoidance consists of a uniform potential pointing from the start to the goal of the desired path, a goal sink potential at the goal of the desired path and several spatial panels surrounding and representing the obstacles.

The strength of the spatial panels must be determined such that the condition on the total potential gradients at the center of the panels are satisfied:

$$V_i = -\frac{\partial \Phi}{\partial n_i}, \quad i = 1..m \tag{3.37}$$

The notation n_i in Eq. (3.37) means that the gradient is in the n_i direction (unit vector perpendicular to the spatial panel) and m is the number of spatial panels. After V_i's are specified, the m Eqs. (3.37) are solved for the unknown panel strength per unit area.

The other condition on the panel strengths (per unit area) is the 3D convergence condition:

$$\lambda_g > \lambda_0 > 0 \tag{3.38}$$

where λ_0 is the total strength of all the spatial panels

$$\lambda_0 = \sum_{i=1}^{m} A_i \lambda_i \tag{3.39}$$

where A_i is the area of panel i.

The total potential field at point $C(x_1, x_2, x_3)$ is:

$$\Phi(x_1, x_2, x_3) = \Phi_u + \Phi_g + \sum_{j=1}^{m} \Phi_j \tag{3.40}$$

where Φ_u (3.22), Φ_g (3.23), Φ_j (3.24) are the potential fields due to the uniform, goal and panel strengths. If Φ'_j defines the potential of the spatial panel j per unit strength per unit area, Eq. (3.40) can be rewritten as:

$$\Phi(x_1, x_2, x_3) = \Phi_u + \Phi_g + \Lambda^T.\Phi' \tag{3.41}$$

where

$$\Phi' = \left(\Phi'_1, \ldots, \Phi'_m\right)^T; \qquad \Lambda = (\lambda_1, \ldots, \lambda_m)^T \tag{3.42}$$

Substituting Eq. (3.41) into Eq. (3.37) results in:

$$V_i = -\frac{\partial \Phi_u}{\partial n_i} - \frac{\partial \Phi_g}{\partial n_i} - \sum_{j=1}^{m} \lambda_j \frac{\partial \Phi_j}{\partial n_i}, \quad i = 1..m \tag{3.43}$$

where

$$\frac{\partial \Phi_u}{\partial n_i} = -\hat{n}_i \left(a_1 \hat{i} + a_2 \hat{j} + a_3 \hat{k}\right) \tag{3.44}$$

$$\frac{\partial \Phi_g}{\partial n_i} = -\hat{n}_i \frac{\lambda_g (x_{c_i} - x_g)}{|x_{ci} - x_g|^3} \tag{3.45}$$

and

$$\frac{\partial \Phi'_j}{\partial n_i} = -\hat{n}_i \left(\frac{\partial \Phi'_j}{\partial x_1}\hat{i} + \frac{\partial \Phi'_j}{\partial x_2}\hat{j} + \frac{\partial \Phi'_j}{\partial x_3}\hat{k}\right) \tag{3.46}$$

The partial derivatives in Eq. (3.46) can be calculated by differentiating the summation term in Eq. (3.24), or approximated numerically. The following relation is presented for the first coordinate:

$$\frac{\partial \Phi'_j}{\partial x_1} \approx \frac{\Phi'_j(x_{ci_1} + \Delta x_{ci_1}, x_{ci_2}, x_{ci_3}) - \Phi'_j(x_{ci_1}, x_{ci_2}, x_{ci_3})}{\Delta x_{ci_1}} \tag{3.47}$$

The same kind of relation is valid for the other two coordinates.

Equation (3.43) is written in the matrix form

$$\boldsymbol{\Lambda} = \mathbf{P}^{-1}(-\mathbf{V} + \Upsilon) \tag{3.48}$$

where

$$\mathbf{P}_{ij} = \left.\frac{\partial \phi'_j}{\partial n_i}\right|_{c_i}, \quad i, j = 1..m \tag{3.49}$$

with

$$\mathbf{V} = (V_1, V_2, \ldots, V_m)^T \tag{3.50}$$

and

$$\Upsilon_i = -\frac{\partial \phi_u}{\partial n_i} - \frac{\partial \phi_g}{\partial n_i}, \quad i = 1..n \tag{3.51}$$

The convergence condition (3.38) must be satisfied:

$$\lambda_0 = \boldsymbol{A}^T \boldsymbol{\Lambda} \tag{3.52}$$

with

$$\boldsymbol{A} = (A_1, A_2, \ldots, A_m)^T \tag{3.53}$$

thus

$$\boldsymbol{A}^T \mathbf{P}^{-1}(-\mathbf{V} + \Upsilon) < -\lambda_g \tag{3.54}$$

The desired outward normal velocities at the center of the spatial panels are assumed to be proportional to the panel area, with a a safety parameter:

$$\mathbf{V} = a\boldsymbol{A} \tag{3.55}$$

Thus

$$a < \frac{\lambda_g + \boldsymbol{A}^T \mathbf{P}^{-1} \Upsilon}{\boldsymbol{A}^T \mathbf{P}^{-1} \boldsymbol{A}} \tag{3.56}$$

with

$$a_{\max} = \frac{\lambda_g + \boldsymbol{A}^T \mathbf{P}^{-1} \Upsilon}{\boldsymbol{A}^T \mathbf{P}^{-1} \boldsymbol{A}} \tag{3.57}$$

and

$$r_a = \frac{a}{a_{\max}} \tag{3.58}$$

The safe λ_j's that satisfy the convergence condition (3.52) can be found from:

$$\boldsymbol{\Lambda} = \mathbf{P}^{-1}\left(-r_a a_{\max}\mathbf{A} + \boldsymbol{\Upsilon}\right) \tag{3.59}$$

Hence, the components of the velocity field are determined as:

$$u_1(x_1, x_2, x_3) = -\frac{\partial \phi_u}{\partial x_1} - \frac{\partial \phi_g}{\partial x_1} - \sum_{j=1}^{m} \lambda_j \frac{\partial \phi_j}{\partial x_1} \tag{3.60}$$

$$u_2(x_1, x_2, x_3) = -\frac{\partial \phi_u}{\partial x_2} - \frac{\partial \phi_g}{\partial x_2} - \sum_{j=1}^{m} \lambda_j \frac{\partial \phi_j}{\partial x_2} \tag{3.61}$$

$$u_3(x_1, x_2, x_3) = -\frac{\partial \phi_u}{\partial x_3} - \frac{\partial \phi_g}{\partial x_3} - \sum_{j=1}^{m} \lambda_j \frac{\partial \phi_j}{\partial x_3} \tag{3.62}$$

3D Robust Harmonic Potential Algorithm

1. The uniform flow strength U, the goal strength λ_g and the safety ratio r_a are selected.
2. The start position $x_s = (x_{s_1}, x_{s_2}, x_{s_3})$ and the goal position $x_g = (x_{g_1}, x_{g_2}, x_{g_3})$ are defined.
3. The 3D obstacles in the environment must be approximated by volumes that can be contained in a number of 3D polygons. The parameters of the polygons are determined such that the vertices starting from an arbitrary vertex in the direction a of right hand notation about the vector $\hat{n}_j$ which is perpendicular to the panel and pointing outward of an obstacle volume are numbered.
4. The direction of the uniform flow is calculated such that the uniform flow points from the start point to the goal point

$$a_1\hat{\mathbf{i}} + a_2\hat{\mathbf{j}} + a_3\hat{\mathbf{k}} = \frac{x_g - x_s}{|x_g - x_s|} \tag{3.63}$$

5. The matrix **P** is evaluated using Eq. (3.49) and Υ by Eq. (3.51)
6. The parameters $a_{\max}$ is calculated using Eq. (3.57)
7. The safety ratio $(0 < r_a < 1)$ is selected.
8. The strength per unit area for the m panels are calculated using Eq. (3.59)
9. Equations (3.60)–(3.62) are used to determine the direction of the local tangent to the path. The path is generated by a numerical procedure, stepping a small constant length along the calculated local tangent. The path starts at $x_1 = x_s$. At any integration step k, the direction of the local tangent to the path and the new position are calculated as:

$$\begin{aligned} u_k &= u(x_k) \\ \hat{u}_k &= \frac{u_k}{|u_k|} \\ x_{k+1} &= x_k + \hat{u}_k \Delta s \end{aligned} \tag{3.64}$$

where Δs is an arbitrary small distance. The iteration in k continues until the point found for the path is closer than the defined small distance Δs

$$\left\| x_{k+1} - x_g \right\| \leq \Delta s \tag{3.65}$$

10. The array of points x_k is the 3D planned path

3.4.3 Sampling Based Trajectory Planning

3.4.3.1 Grid Based State Space Search

This method defines an arbitrary speed-varying safety corridor, making this particular algorithm one of very few trajectory planning algorithms with a proven explicit safety guarantee. This algorithm discretizes the entire state space of the vehicle onto a lattice, and searches the lattice for the time-optimal path that satisfies the safety corridor. Although the algorithm converges as a polynomial of the number of obstacles, it is a high-order polynomial that is exponential with the number of dimensions, making practical real-time implementation difficult due to high dimensionality of the state-space. It has also difficulty in solving planning problems for the case of under-actuated vehicles, as the lighter than air robot.

3.4.3.2 State Space Navigation Function with Interpolation

Much like the grid-based state space search, this method approximates the time-optimal path to a goal. Instead of returning only a single trajectory, it returns a navigation function over the state space. This navigation function can be computed by either value iteration or control policy iteration, although value iteration is more popular. For a given state, performing gradient descent on this navigation function will produce an approximately time-optimal trajectory to the goal. Interpolation between lattice points allows a continuous function which can be used for feedback. The algorithm takes on the same order of complexity as the grid-based state space search.

3.4.3.3 Reachability Graph

This approach also uses a body-centered frame of reference: for each state, the tree explores variety of states including the maximum deflection states. The combinatorial complexity of such a process is often prohibitive, and the tree quickly fills the space close to the initialization point. The basic approach has been employed for curvature constrained path problem in 2D. Another way to make this approach tractable is to use cell-based pruning: the configuration space is divided into cells, and the reachability graph is set up to be a tree that has no more than one leaf ending in each cell.

3.4.4 *Decoupled Trajectory Planning*

3.4.4.1 Minimum Distance Discrete Path Followed by Trajectory Forming (2 Step Approach)

This algorithm follows the same general approach: first a discrete path through the configuration space is found by one of the above presented algorithms, and then the resulting path is used as the basis for the generation of a trajectory that is feasible for the dynamics-constrained vehicle. For the first stage, a well-known algorithm such as A*, the probabilistic road map, or the Voronoi approach is typically used. Complexity in this approach is dominated by the path planning phase computations. This decomposition-based approach allows efficient computation of approximate solutions, but makes proofs of completeness, optimality, or even of soundness, difficult. There is no safety corridor built in and soundness requires checking whether the trajectory intersects with obstacle space, and re-computing or rejecting trajectories that do not pass. Generally, practical implementations of these algorithms will use a conservative corridor to prevent collisions.

3.4.4.2 Discrete C-Space Search Connected by 2 Point Free Space Boundary Value Solver

In this approach, a set of waypoints is first selected (generally by a grid based search), a velocity is assigned to each one, and a boundary value problem connecting each way point to the next point is solved. Although these boundary value problems do not have to deal with obstacles, a general simple solution is not possible, so the solution is generally approximated using numerical methods. As with the previous method, an explicit collision check on the trajectory is needed to ensure soundness, possibly at the expense of completeness.

3.4.4.3 Discrete C-Space Search Interpolated with Polynomial Arcs

In this approach, an ordered set of waypoints produced by a discrete planner is fitted with a spline made up of polynomial arcs. This spline is set up so that the vehicle can follow it without violating acceleration constraints, and typically consists of circular arc segments with a minimum radius and straight segments [69]. Other types of spline, such as the Pythagorean hodograph have been proposed for the same purpose, in 2D [68, 91].

The Curvature Constrained Paths with Speed Control Planning is coupled with the previous approach; once a curvature constrained path is produced, it is possible to optimize the speed so that the vehicle will follow this path in minimal time. This path-constrained trajectory planning problem requires solving only a 2D path constrained state space (with time and velocity axes) and can be accomplished efficiently.

3.4.4.4 2D Voronoi Solutions from Multiple Body-Based Planes

In this approach, several planes containing both the initial and final goal points are extracted as subsets of the 3D configuration space. These planes that are discriminated from each other by a single angle and this angle is discretized so there are a finite number of planes (in general, 4 to 6 planes suffices). These planes are then searched with a 2D dimensional path planner, such as Voronoi roadmap planner, and then rated against each other according to optimality criteria. The best of the set according to the performance index, is chosen.

3.4.5 The Finite State Motion Model: The Maneuver Automaton

The general idea of finite state models is to reduce the optimization or search problem from an infinite dimensional space to a finite one. It helps to significantly reduce the computational complexity of a trajectory optimization [71, 72].

There are two primary types of finite-state models for dynamics systems: the first is a discrete-time model with quantized states (quantization); another choice is to relax the restrictions on control and time and use instead operations over discrete stages with fixed start and end states. These stages, which are feasible time-parameterized curves in state space, are called motion primitives. In the context of vehicle trajectory planning, this model is called a maneuver automaton (MA). The concept of Maneuver Automaton for vehicle is based on the observation that human pilots achieve control using a combination of trim trajectories and maneuvers (non-equilibrium transitions between trims). Frazzoli et al. [72] provides a definition of the concept of Maneuver Automaton within the context of autonomous guidance. Maneuver Automaton form the set of trim and maneuvers are used to pre-compute a cost-to-go map. This map can be used on line with a greedy guidance policy. States falling between the pre-computed values are obtained via interpolation. In this form, the vehicle behavior is constrained to the set of primitives in the maneuver automaton.

Remark 3.6 For agile vehicles such as robotic helicopters, it can be an issue to achieve a sufficiently expressive maneuver automaton due to the 'curse of dimensionality'.

To relax the vehicle behavior and provide flexibility in an obstacle-rich environment, Schouwenaars et al. in [184] use the concept of the Maneuver Automaton within a receding horizon optimization framework. Instead of fixed trim trajectories, the trims are replaced by controllable linear modes.

Parallel to the Maneuver Automaton concept, a similar idea called 'control quanta' is introduced for driftless systems with a symmetry property in [142]. For this special class of systems, by employing control quanta, the reachable set can be restricted to a lattice, and by choosing a suitable set of control quanta, the reachable set can be everywhere dense in the limit when M approaches infinity. The difference is that for the control quanta method, the control policy is chosen from a collection of control library policies, while for motion primitive method the trajectory is chosen from a library of maneuvers that can result from a various control strategies.

3.4.6 Mathematical Programming

Mathematical programming methods treat the trajectory planning problem as a numerical optimization problem. Some popular methods include Mixed Integer Linear Programming (MILP), non linear programming and other constrained optimization approaches. These methods are also known as trajectory optimization methods, since they find a trajectory to a goal point that is optimal in the resolution sense. However, the cost functions typically have a number of local minima thus finding the global solution strongly depends on the initial guess (the general formulation is NP hard, although given an initial guess sufficiently close to the global solution, the optimization converges in polynomial time). For this type of problem, one standard strategy is to enforce the equations of motion as constraints. Another strategy is to discretize the variational principles underlying the systems dynamics, such as Hamilton's principle or Lagrange-D'Alembert principles, and then these discrete equations can serve as constraints. This kind of strategy is called Discrete Mechanics and Optimal Control (DMOC). Several approaches have been used to break this into simpler problems.

The MILP mixed integer linear programming approach [178] uses indirect branch-and-bound optimization, reformulating the problem in a linearized form and using commercial software to solve the MILP problem. There are two elements to a generic brand-and-bound method [59]

- Branching: The problem is divided into subproblems by partitioning the search space. Each subproblem is further divided in the same way, and the algorithm proceeds by searching a tree of sub-problems, hence branching
- Bounding: a lower bound on the optimal cost of each subproblem is found by solving a relaxed simpler form of that subproblem. These bounds are then used to identify if a branch requires further subdivision. If the subproblem is infeasible or the solution to the subproblem is worse than the cost of the best feasible solution found so far, then the branch is said to be fathomed and no further branching is necessary.

The following discussion illustrates the direct mapping of the branch and bound to the 2D avoidance problem.

- Branching: the problem of avoidance obstacles 1..M can be divided into two subproblems
  - the problem of passing clockwise around obstacle 1 and avoiding obstacles 2..M or

- the problem of passing counterclockwise around obstacle 1 and avoiding obstacles 2..M
- Bounding: the shortest path passing clockwise (or counterclockwise) around obstacle 1 and avoiding obstacles 2..M must be longer than the shortest path passing clockwise (or counterclockwise) around obstacle 1 and ignoring obstacles 2..M

The global optimality of any solution obtained from a branch and bound method is guaranteed, assuming that the globally optimal solution to each evaluated subproblem is found. In the avoidance case, this corresponds to finding the best path passing on a given side of a set of active obstacles and ignoring all others. Because this subproblem no longer involves a choice of side, it can be solved by nonlinear programming. Although, it is difficult to express the concept of one side or another in terms of an explicit constraint, it can be achieved by initializing the search on the appropriate side and employing a primal-dual optimization method which is then checked against the desired path.

An initial trajectory for the mathematic programming methods, such as a constant-speed trajectory, is then used as an initial point in the mathematical programming search. If this initial point falls within the basin of attraction of the global solution, then the mathematical programming approach can find the optimal solution in polynomial time. However, unless care is taken in finding proper initial points, the solution could fall into a local minimum, and general global optimization approaches guaranteed to find the global minimum are prohibitively expensive. Mixed Integer Linear Programming can extend continuous linear programming to include binary or integer decision variables to encode logical constraints and discrete decisions together with the continuous vehicle dynamics. The approach to optimal path planning based on MILP was introduced in [30, 136, 184] for robotic helicopters. The UAV's trajectory generation is formulated as a 3D optimization problem under certain conditions in the Euclidean space, characterized by a set of decision variables, a set of constraints and the objective function. The decision variables are the UAV's state variables, i.e. position and speed. The constraints are derived from a simplified model of the UAV and the environment where it has to fly on. These constraints include:

- Dynamics constraints, such as a maximum turning force which causes a minimum turning radius, as well as a maximum climbing rate.
- Obstacles avoidance constraints like no-flight zones.
- Target reaching constraints of a specific way point or target.

The objective function includes different measures of the quality in the solution of this problem, although the most important criterion is the minimization of the total flying time to reach the target.

3.4.7 Receding Horizon Control

Receding Horizon Control (RHC) solves the numerical optimization problem over a reduced time horizon. In this approach, an open-loop control policy is designed to control the vehicle until the end of the time horizon. Optimization over a finite horizon requires reduced computation time, however, it will not converge to a globally optimal solution without using an appropriate cost-to-go function to capture the discarded portion of the trajectory. Except in trivial cases, optimality and completeness are difficult to prove. In helicopter robots guidance applications, this approach has often been used with a MILP solution.

Receding Horizon Control is a suitable trajectory planning technique for many UAV applications. Sensory information can be incorporated into on-line computation thus it can deal with uncertainty; at the same time, only local information is integrated thus it can reduce computational effort. However, properly designed terminal cost function needs to be provided to the on-line planner to guarantee completeness and near-optimality. The value function captures the relationship between the vehicle dynamics (state), the environment and the cost.

3.4.8 Reactive Planning

The term 'reactive planning' refers in general to a broad class of algorithms that use only local knowledge of the obstacle field to plan the trajectory. Reactive algorithms are important in dealing with uncertainty, and run very quickly since no elaborate couplings are involved. In the case where a global obstacle map is not available and obstacle positions are known within a small radius, a reactive algorithm prevents last-minute collisions by stopping or swerving the vehicle when an obstacle is known to be in the trajectory, which has been planned by a different algorithm. This type of approach is important in many existing practical implementations in order to 'patch' an unsound algorithm to ensure that it is sound, as well as to deal with obstacle fields that may change suddenly. However, reactive planners, due to their inability to take the global planning problem into consideration, are seldom used as the sole trajectory generation process. If only the reactive planner is used, the vehicle may never find a trajectory that will lead to the goal.

The Motion Description Language (MDL) and its extension (MDLe) can be used to define and design the reactive algorithms. In this framework, a sensor based interrupt (i.e. obstacle detected) will cause the vehicle to switch to another behavior. Fuzzy logic could also be used as a basis for a reactive algorithm, as well as visual flow based reactive planning.

There are many ongoing works on this topic, especially bio-inspired reactive planning algorithms. This can be explained due to the deficiencies of existing planning algorithms to involve sensory information in a principled way so that a complete planning framework results.

3.4.9 Probabilistic Roadmap Methods: PRM

The roadmap methods described above are able to find a shortest path on a given path. The issue most path planning methods are dealing with is how to create such a graph. To be useful for path planning applications, the roadmap should represent the connectivity of the free configuration space well and cover the space such that any query configuration can be easily connected to the roadmap. PRM is a probabilistic complete method that is able to solve to solve complicated path planning problems in arbitrarily high dimension configuration spaces.

Probabilistic planners represent a class of efficient methods. They belong to the general family of sampling-based methods. A configuration that does not cause a collision is added to the current road map and connected if possible to other already stored configurations.

The above strategy is quite general and may lead to different planning methods depending on the specific design choices, and mainly on the criterion for selecting the samples in to be checked for collision. The basic PRM Approach constructs a roadmap by iteratively sampling configurations randomly from the configuration space. Using a collision checker, it can be determined whether a configuration belongs to the free or forbidden configuration space. If a configuration is collision-free, it is added as a node to the roadmap. Subsequently, it is attempted to connect this node to the roadmap. To save time, a connection is only tried to nodes which are close. The set of nodes to which a connection is attempted is called the neighbor set of the current node.

Connections between nodes are tried using a local planner which is a simple planner that is allowed to fail on all but the simplest queries. In The basic PRM Approach, the local planner simply tries to connect two configurations by a straight line through the configuration space. The local planner succeeds when the straight line is collision free, which is collisions checking intermediate configurations, up to some predefined resolution. If the connection succeeds, an edge between the two associated nodes is added to the roadmap.If a connection has been tried to all nodes of the neighbor set, a new node is sampled, and the process repeats. This continues until some application specific stop criterion is met. Usually, this is when some predefined set of query configurations are inter-connected via the roadmap.

The local planner should be able to find a path between two configurations in simple cases in a small amount of time. Given a configuration and a local planner, one can define the set of configurations to which a local planning attempt will succeed. This set is called the visibility region of a node under a certain local planner. The larger the visibility region is, the more powerful the local planner.

The most straightforward sampling scheme is to sample configurations uniform randomly over the configuration space.

3.4.10 Rapidly Expanding Random Tree (RRT)

The Rapidly Expanding Random Tree is an algorithm which is suited for quickly searching high-dimensional spaces that have both algebraic and differential constraints. The key idea is to bias the exploration toward unexplored portions of the space by sampling points in the state space and incrementally pulling the search tree toward them, leading to quick and uniform exploration of even high-dimensional state spaces. The underlying premise is to build a graph structure with nodes at explored positions and with edges describing the control inputs needed to move from node to node [111].

For each step, a random state (x_{rand}) is chosen in the state space. Then (x_{near}) in the tree that is the closest to the (x_{rand}) in metric ρ is selected. Inputs $u \in \mathbf{U}$, the input set are applied for Δt, making motions toward (x_{rand}) from (x_{near}). Among the potential new states, the state that is as close as possible to (x_{rand}) is selected as a new state (x_{new}). The new state is added to the tree as a new vertex. This process is continued until (x_{new}) reaches (x_{goal}).

Since a vertex with a larger Voronoi region has a higher probability to be chosen as (x_{near}) and it is pulled to the randomly chosen state as close as possible, the size of larger Voronoi regions is reduced as the tree grows. Therefore, the graph explores the state space uniformly and quickly. To improve the performance of the RRT, several techniques have been proposed such as biased sampling and reducing metric sensitivity [111].

This method uses a stochastic search over the body-centered frame of reference and expands a tree through a random sampling of the configuration space. This algorithm is proven to be complete in the probabilistic sense, and to produce a trajectory that is feasible given the dynamic constraints of the vehicle. However, there is no proof of the convergence rate or of optimality.

3.4.11 Guided Expansive Search Trees

Guided Expansive Search Trees is a variation of conventional probabilistic path planning strategies: PRM, RRT and most similarly Expansive Search Trees (EST). Guided Expansive Search Trees have advantages over the conventional techniques when the state space is governed by higher dimensions, Kinodynamic constraints and when minimizing the control cost of the entire path is important. Guided Expansive Search Trees borrow from conventional expansion techniques for robustness in finding paths and also use path cost statistics to guide the tree in the window of acceptably low-cost paths [169].

Conventional probabilistic path planning algorithm rely on a metric to determine whether two configurations are 'close'. But in kinodynamic state spaces, it is not obvious when two configurations should be considered 'close'. Often, a 1-norm or weighted 1-norm of the vector of the configuration is used in a kd-tree or range tree, or only the 1-norm of the static representation is used, ignoring the derivatives and time. These techniques do not necessarily accurately represent the reachability of the configuration. For instance, two configurations which are close based on a 1-norm will likely not be able to reach one another with a reasonable cost.

The pseudo-code for this algorithm is the following

1. For i = 0 to N do
2. p = choose-waypoint()

3. n = expand-waypoint(p)
4. (if n is not valid) continue
5. add-to-tree (p, n)
6. assign-weigh (n)
7. if n connects to goal: return add-to-tree (n, goal)
8. end for.

The weighting function of the Expansive Search Trees algorithm only looks at the number of neighbors n_{neigh} within a range $weight = \frac{1}{n_{neigh}}$.

The guided expansive search tree algorithm differs in that it additionally takes into account the out-degree, number of out-going edges from the waypoint; the order of the waypoint; how recently it was created and the A* cost, estimated total cost to the goal computed as the sum of the control required to reach the waypoint from the root and the estimated control cost to reach the goal. These statistics are taken to the power α, β, γ, δ respectively.

$$weight = \frac{(order)^{\gamma}}{n_{neigh}^{\alpha}(out - degree)^{\beta}(A^{*}\ cost)^{\delta}}$$

The out-degree term prevents a highly weighted waypoints from being expanded too many times. This is incremented even if the expanded to waypoint is not valid. The A* cost term focuses the search towards the goal and prevents the tree from often expanding high-cost waypoints which violate a velocity or rate constraints. The order term, like the number of neighbors term, tends to keep the tree expanding on the frontier.

The purpose of using the number of neighboring waypoints in the weighting function is to bias the search towards expanding on the frontier. So, whether two waypoints are close should be defined based on the control cost between the two configurations. This indicate how likely one way-point is to expand into the region of the other.

3.5 Planning with Uncertain Winds

One of the most challenging applications is planning under uncertainty. A flight planning algorithm that is robust to changing flight conditions, disturbances and vehicle uncertainties would be an improvement over current flight technology. The general problem of planning with uncertainty can be phrased as follows: Given a vehicle with uncertain position information, uncertain environment knowledge (e.g. obstacle locations) and having limited precision in tracking commands, find the best path to the goal. In most of the real world UAV planning problems, the issue of uncertainty in sensing and control is unavoidable. The Receding Horizon Control framework described earlier can also be employed to deal with uncertainties; the off-line pre-planned trajectory or approximate cost-to-go function accounts for global convergence and the on-line Receding Horizon Control can be used to negotiate with mid-flight uncertainties. It can be used to generate trajectories for a vehicle operating in an environment with atmospheric turbulence.

Remark 3.7 When the actual wind vector field is not known exactly and may deviate significantly from the wind velocities estimated by the model, a technique to address this issue is to explicitly incorporates wind uncertainties into the planning algorithm.

When systematic search methods fail, less conventional search techniques have to be considered. Many of them employ some form of random or stochastic search, including Receding horizon control, Markov decision process and chance constrained algorithms. Heuristic methods are techniques which seek good solutions at reasonable computational cost without being able to guarantee optimality.

Remark 3.8 The field of Simultaneous Localization and Mapping (SLAM) is important in problems of planning with uncertainty. A common line of attack for solving such problems in real time is to break the problem into two phases: an off-line phase and an on-line phase. The off line phase consists of solving the optimal control problem for various reference trajectories and storing these reference trajectories on board for later on line use. These reference trajectories are used to compute the actual trajectory on line via a neighboring optimal feedback control strategy, typically based on the linearized dynamics. This approach requires extensive ground based analysis and on board storage capabilities. Moreover, perturbations around the reference trajectory might not be small and therefore applying the linearized equations may not be appropriate. Although the SLAM problems is closely linked with the problem of motion planning with uncertainty, it does not immediately address the motion planning issues, so it is not covered in this book.

3.5.1 Receding Horizon Approach

For manned aircrafts, the nominal path is specified by the air traffic controller by a sequence of way-points and is typically a piecewise linear one. The coordinates of the way-points are assumed to be given in a global coordinate frame. The speed is typically measured in nautical miles per hour. Aircraft motions are subject to various random perturbations such as wind, air turbulence ... and thus may deviate from the nominal path. This cross-track deviation may be corrected by the on board Flight Management System (FMS). In addition, aircraft dynamics may exhibit several distinct modes, for example, keeping constant heading, turning, ascending, descending and may switch modes at proper times when following the nominal paths. The aircraft is assigned some flight plan to follow that consists of an ordered sequence of way-points $\{P_i, i = 0 \dots M\}$, $P_i = (x_i, y_i) \in R^2$. Ideally, the aircraft should fly at some constant speed along the reference path composed of the concatenation of the ordered sequence $\{I_i, i = 1 \dots M\}$ of line segments I_i, with starting point P_i and ending point P_{i+1}, $i = 1 \dots M$. Deviations from the reference path may be caused by the wind affecting the aircraft position and by limitations in the aircraft dynamics in performing sharp turns resulting in cross-track error. The on board 3D FMS tries to reduce the cross-track error by issuing corrective actions based on the aircraft's current geometric deviation from the nominal path, however, without taking into account timing specifications. The obtained results are applied to a clearance changing the flight plan. The position of the aircraft may be measured via secondary radar or other localization systems such as Automatic Dependent Surveillance Broadcast (ADS-B) [86]. These measurements are precise enough to neglect measurement noise. Current methods of position localization produce errors of order of terms of meters, which are small compared to the effect of other sources of nominal uncertainty such as winds. The flight plan possibly involves altitude changes. Altitude changes can be used as resolution maneuvers to avoid severe weather areas or other conflict situations with other aircraft. Forbidden airspace areas may have an arbitrary shape, which can also change in time, as for example, in the case of a storm that covers an area of irregular shape and evolves dynamically.

As for the manned aircraft, the evolution of the lighter than air robot is subject to many natural sources of uncertainty, roughly classified into two classes.

- Nominal uncertainty, that affects all flights, results of generic perturbations due to weather, variability in the mass of the aerial robot, variability in the settings of the Mission planning or control systems, biased measurements... Nominal uncertainty gives rise to quantitative differences between the actual lighter than air robot and the model used to predict its evolution.
- Non nominal uncertainty that affects certain flights, result of things as malfunctions, errors of the human operators, extreme weather conditions. Non nominal uncertainty gives rise to qualitative structural differences between the actual air traffic and the model used to predict its evolution.

3.5.1.1 Basic Problem Statement

In a receding horizon approach, a trajectory that optimizes the cost function over a period of time, called the planning horizon, is designed first. The trajectory is implemented over the shorter execution time and the optimization is performed again starting from the state that is reached at the end of the execution time. The terminal condition is assumed to be changing with time.In this section, the idea is simple: as uncertainty always exists, a way of taking care of it is to update the reference paths either periodically or when necessary. The immediate measurements of the position and orientation are taken as initial conditions for the next reference trajectories [20].

When the target point is far off then there is no real advantage to finding the optimal trajectory on line with high precision from the starting point until the end. As the lighter than air robot continues to move to the target, the flight management system can get more accurate information about the surrounding environment (path constraints) which may be different from what was assumed at the beginning when the trajectory was optimized. Moreover, the path constraints change in a dynamic environment.

3.5.1.2 Hierarchical Planning Structure

This approach involves determining the optimal open loop motion plans based on the available information and world model and repeating this process to refine the plan as the information becomes more accurate [14]. The structure necessary for the update of the reference trajectories is described in this paragraph. A hierarchical structure appears as there exists an upper level of decision making and a lower level where the motion planning problem is solved. One clock, associated with the lower level, allows the reference trajectories design, depending on the nature of the next way-point. It is reset at each update. The following concepts must be introduced to describe this hierarchical structure operation

- Periodic Updates: First, periodic updates are introduced with a period 10 to 1000 times greater than the step of the autopilot.
- Anticipated Updates: An important perturbation may occur between two periodic updates. To handle this situation, a new reference trajectory is required. The concept of anticipated update is thus important. To decide whether an important perturbation occurs the system needs a supervision level.
- Supervision level: The supervision system is based on a spatio-temporal criterion.
  - The temporal part: It depends on the value of a parameter called e_{ct} computed continuously in the lower level. The parameter e_{ct} gives an evaluation of the arrival time predicted in the kth update and the arrival time predicted continuously on the basis of the measured configuration, T_c:

$$e_{ct} = |T^k - T_c - h| \tag{3.66}$$

    where h is the time given by the clock. If $e_{ct} > \varepsilon_{\min}$ then an update occurs; $\varepsilon_{\min}$ is user-fixed relatively to the environment
  - The spatial part: The spatial parameter is defined as $e_s = |X_r - X|$ where X_r is the reference configuration and X the measured one. If $e_x > \delta_{\min}$ then an update must take place; The parameters $\varepsilon_{\min}$ as $\delta_{\min}$ influence the number of updates.

The above procedure has been developed to be implemented on embedded systems aboard the lighter than air robot [1, 14, 27]. Periodically updated paths can be generated to accommodate a slowly drifting wind direction and/or wind speed. A flight planning system is presented that enables the vehicle to move in a non-predefined way. Taking into account the situation detected by the sensors and on the basis of a spatio-temporal criterion, the proposed system is able to decide whether a new reference trajectory must be planned. If an update is necessary, the reference is computed taking into account the measured states, the limitations on thrust and velocity and the next way-point (or final destination). It is assumed that the sampling interval is small enough that the discretization effects can be neglected.

3.5.2 Markov Decision Process Approach

To prepare the motion planning problem, the uncertainty in the wind field is modeled. The problem of reaching a particular goal location as a Markov Decision Process (MDP) is formulated using a discretized space approach. Solving the Markov Decision Process provides a policy of what actuation option should be expected at any given location.

3.5.2.1 Introduction

A particular finite Markov Decision process is defined by its state and action sets and by the one-step dynamics of the environment. Given any state s and action a, the probability of each possible next state s', is

$$P^a_{ss'} = \Pr\left\{s_{t+1} = s' \| s_t = s, a_t = a\right\}$$

where $P^a_{ss'}$ represents transition probabilities and t denotes a finite time step [104]. In the Markov Decision Process, the value of $P^a_{ss'}$ does not depend on the past state transition history. The agent receives a reward r every time it carries out the one-step action. Given any current state s and action a, together with any next state s′, the expected value of the next reward is:

$$R^a_{ss'} = E\left\{r_{t+1} \| s_r = s, a_t = a, s_{t+1} = s'\right\}$$

$P^a_{ss'}$ and $R^a_{ss'}$ completely specify the dynamics of the finite Markov Decision Process. In the finite Markov Decision Process, the agent follows the policy Π. The policy Π is a mapping from each state s and action a to the probability $\Pi(s,a)$ of taking action a when in state s. In the stochastic planning calculation, based on the Markov Decision Process, the policy Π is decided so as to maximize the value function $V^{\Pi}(s)$. The $V^{\Pi}(s)$ denotes the expected return when starting in S and following Π thereafter. The definition of $V^{\Pi}(s)$ is:

$$V^{\Pi}(s) = E_{\Pi}\left[\sum_{k=0}^{\infty} \gamma^k r_{t+k+1} \middle\| s_t = s\right]$$

where E_{Π} denotes the expected value given when the agent follows the policy Π and γ is the discount rate $0 < \gamma < 1$. If the values of $P^a_{ss'}$ and $R^a_{ss'}$ are known, dynamic programming is used to calculate the best policy Π that maximizes the value function $V^{\Pi}(s)$. When the values of $P^a_{ss'}$ and $R^a_{ss'}$ are unknown, a method such as on-line reinforcement learning is useful in obtaining the best policy Π in the learning environment. After the planning calculation has finished, a greedy policy that selects action value a that maximizes $V^{\Pi}(s)$ is optimal.

Because the wind velocity $W(r_i,t)$ is uncertain, the next state s_j may be considered as a random variable and a probability distribution can be constructed over all adjacent cells. Given these transition probabilities from all states, the motion planning problem is to select the actions that minimizes time-to-goal. This problem is thus naturally posed as a Markov Decision Process (S, A, P, R) where S represents the set of possible states s_i, A is the set of actions available from each state, P gives the transition probabilities $P_a(s_i, s'_j)$ from current state s_i to possible next states s'_j under action a, R defines the expected immediate reward for each transition and each transition a. In this problem, reward is negative travel time.

3.5.2.2 Modeling Wind Uncertainty

The wind is decomposed into direction and magnitude components, denoted respectively by θ_i and W_i. Independent probability distributions are assigned to each. The expected value of each component is equal to the field model value given by $\bar{W}_i(r, t)$. To model the uncertainty in θ_i, a Von Mises distribution is employed

$$f_{vm}\left(\theta_i, \bar{\theta}_i\right) = \frac{\exp(k\cos(\theta_i - \bar{\theta}_i))}{2\pi I_0(k)} \tag{3.67}$$

where $\bar{\theta}_i = \angle\bar{W}_i(r, t)$ is the mean, k is a concentration parameter and $I_0(k)$ is the modified Bessel function of the first kind of order 0. The wind magnitude W_i uncertainty is modeled as Gaussian with mean $\bar{W}_i = \|W(r_i, t)\|$. The standard deviation is set as proportional to the magnitude, $\sigma_i = \rho W_i$ (ρ is chosen by the user).

3.5.2.3 Transition Probabilities

Next, the transition probabilities $P_a(s_i, s'_j)$ are defined that govern state s'_j is entered after executing each action a from each state s_i. If there is no vehicle actuation, the wind direction wholly determines the next state's position. In that case, the probability of the wind forcing the vehicle from state s_i to an adjacent state s_j is

$$P_a(s_i, s'_j) = \int_{\theta_{ij}}^{\theta_{ij}+\pi/4} f_{vm}\left(\theta_i, \bar{\theta}_i\right) d\theta \tag{3.68}$$

where θ_{ij} is the smaller bordering angle of the $1/8$ circular sector pointing from s_i to s_j.

A Monte Carlo method may be used to determine the probabilities P and expected rewards R from resultant vector for each action, described as follows. First a set θ_0 of N sample points are drawn from the Von Mises distribution $f_{vm}(\theta, \bar{\theta})$ and a set W_0 of N sample from the standard normal distribution (this step is required only once, whereas the following steps must be done by iterating for each s_i). Second these samples are adjusted for the state s_i. Third, the resultant velocity samples are calculated by converting the random wind samples to Cartesian coordinates and adding the actuation.

3.5.2.4 Action Space and Rewards

Travel time is used as a transition cost between states. The expected value of the velocity magnitude is estimated as the population mean of the above Monte-Carlo samples.

To complete the problem set-up, a goal location is defined and created as a sink state at this location, from which all transitions are set to have probability zero. Thus the cumulative reward will decrease with every transition until the vehicle reaches the goal location.

Given this set-up, the Markov Decision Process will determine for each given current state s_i what is the optimal immediate action a so that the expected cumulative time to goal is minimal, this collection of actions is referred to as the optimal policy Π^*.

3.5.3 *Chance Constrained Predictive Control Under Stochastic Uncertainty*

Lighter than air robots must be able to plan control actions that are robust to the inherent uncertainty in the real world. This uncertainty arises due to uncertain state estimation, disturbances and modeling errors, as well as stochastic mode transitions such as component failures. Some authors [32] have investigated control under set-bounded uncertainty. In this case, robust control ensures that failure is prevented under all possible uncertainties. In many cases, for example wind disturbances, uncertainty is best represented using a stochastic model, rather than a set bounded one. Early approaches of control under stochastic uncertainty, such as linear Gaussian quadratic used the certainty equivalence principle. This enables uncertain variables to be replaced with their expectation and control laws to be designed in terms of these expectations. Alternative approaches expressed uncertain classes of stochastic control problems as a Markov decision process. Predictive stochastic control takes into account probabilistic uncertainty in dynamic systems and aims to control the predicted distribution of the system state in some optimal manner over a finite planning horizon. Chance constraints specify that the probability of failure must be below a given threshold. Failure can be defined as collision either with an obstacle or failure to reach the goal region. This chance constrained formulation enables the user to specify a desired level of conservatism, which can be traded against performance. It is much beyond the calculation possibilities of the actual embedded systems, but technology is advancing fast.

The key idea is to approximate all probability distributions using a finite set of samples or particles. The stochastic predictive problem is approximated as a deterministic one. As the number of particles tend to infinity, the approximation tends to the original stochastic problem. Constraints are incorporated on the probability of failure, enabling chance constrained stochastic control with continuous decision variables. The proposed formulation based on the analysis presented in [32].

3.5.3.1 Problem Formulation

The following definitions are taken: the states $X = (x, y, z, \gamma, \chi, V)$, the controls $U = (T, \sigma, \alpha)$ the model parameters $\Theta = (C_D, C_L, m, m_{ax}, m_{ay}, m_{az})$ and finally disturbances $\mathbf{W} = (W_x, W_y, W_z, \dot{W}_x, \dot{W}_y, \dot{W}_z)$.

The states of the system are defined by the following function

$$\begin{aligned} X_1 &= f_1(X_0, U_0, \Theta_0, \nu_0) \\ X_2 &= f_2(X_0, U_0, U_1, \Theta_0, \Theta_1, \nu_0, \nu_1) \\ &\dots \\ X_T &= f_T(X_0, U_{0:T-1}, \Theta_{0:T-1}, \nu_{0:T-1}) \end{aligned} \tag{3.69}$$

X_T is used to denote the value of variable X at time T, and $X_{1:T}$ denotes the sequence $\langle X_1, X_2, \dots, X_T \rangle$. The initial state, model parameters and disturbances are modeled as random variables. The problem is to design a finite, optimal sequence of control inputs, taking into account probabilistic uncertainty, which ensures that the state of the system leaves a given feasible region $\mathbf{F}$ with probability at most ε and keeps the expected state within another feasible region $\mathbf{F}$. This can be formulated as:

Problem 3.1 Minimize $E[h(U_{0:T-1}, X_{1:T})]$

Subject to

$$\begin{aligned} p(X_{1:T} \notin \mathbf{F}) &\le \varepsilon \\ E(X_{1:T}) &\in \mathbf{G} \\ U_{0:T-1} &\in \mathbf{U} \end{aligned}$$

where h is a cost function defined over the control inputs and system state trajectory. $\mathbf{F}$ is an operator defined feasible region for the system state trajectory, $\mathbf{G}$ is an operator defined feasible region for the expected state trajectory and $\mathbf{U}$ is an operator defined feasible region for the control inputs.

For the lighter than air robot planning, $\mathbf{F}$ is defined so that the system state is in the goal region at the final time step and avoids a number of obstacles at all time steps. $\mathbf{G}$ is defined so that the expected velocity is identically zero at the final time. Optimality can be defined in terms of either minimizing control effort or time to reach the goal... Wind and turbulence disturbances are considered.

Weighting factors are needed for cost functions that have more than one term. The shortest path that uses the least amount of fuel is often neither the shortest possible path, nor the path that uses the least fuel, but one which strikes a balance between them. The relative weights of these terms determine what sort of balance results. Substantial oversight is often required to analyze the sensitivity of solution characteristic to cost function weights and this sensitivity analysis may be specific to particular problems rather than fully generalizable.

Changing the weights changes the resulting behaviors. If saving fuel and avoiding obstacles are highly prized, the robot may travel slowly, avoid rapid accelerations and looping far around obstacles. Behaviors that are optimal with respect to specific physical quantities (e.g. fuel, time and distance from obstacles) are sought. Given the flight management system goals and environment, some of the quantities may be perceived as more important than others and weighted more heavily, resulting in a behavior that is optimal for this LTAR, with these goals, in this particular environment. A computational modeling system is presented, that recognize what kind of behaviors is mostly likely to be successful, given the goal and the environment at the moment.

3.5.3.2 Algorithm

Approximating the probability distribution of a random variable using samples (or particles) can lead to tractable algorithms for estimation and control. A particle represents a state trajectory over the entire planning horizon.

This method works by approximating all probabilistic distributions with particles, thereby approximating an intractable stochastic optimization as a tractable deterministic optimization problem. By solving this deterministic problem, an approximate solution to the original stochastic problem is obtained, with the additional property that as the number of particles used tends to infinity, the approximation becomes exact.

Algorithm Chance constrained particle control algorithm

1. Generate N samples from the proposal distribution $q(X_0, \mathbf{W}_{0:T-1}, \theta_{0:T-1})$ defined over the initial state, disturbances and model parameters.
2. Calculate the importance weight ω_i for each sample

$$\omega_i = \frac{p(X^{(i)})}{q(X^{(i)})}$$

3. Express the distribution of the future state trajectories approximately as a set of N particles; each particle $X_{1:T}$ corresponds to the state trajectory given a particular set of samples $\{X_0^{(i)}, \nu_{0;T-1}(i), \Theta_{0:T-1}(i)\}$ and depends explicitly on the control inputs $U_{0;T-}$ which are yet to be generated.
4. Approximate the chance constraints in terms of the generated particles. The probability of failure is approximated as follows $P(X_{1:T} \notin F) \approx \frac{1}{N}\sum_{i=1}^{N} \omega_i g(X_{1:T}^{(i)})$ where $P_A = E_X[g(x)]$ with

$$g(X) = \begin{cases} 0 & f(X) \in A \\ 1 & f(X) \notin A \end{cases}$$

The approximated chance constraint is then $\frac{1}{N}\sum_{i=1}^{N} \omega_i g(X_{1:T}^{(i)}) \leq \delta_0$.
5. Approximate the constraints on the expected state using the sample mean approximation $\frac{1}{N}\sum_{i=1}^{N} \omega_i X_{1:T}^{(i)}$.
6. Approximate the expected cost in terms of particles: $\hat{h} = \frac{1}{N}\sum_{i=1}^{N} \omega_i h(U_{0\ldots T-1}, X_{1:T}^{(i)})$. A weighed fraction of no more than δ_0 of the particle can fall outside of the feasible region.

7. Solve deterministic constrained optimization problem minimizing $\hat{h}$ over control inputs $U_{0...T-1} \in U$ subject to $\frac{1}{N}\sum_{i=1}^{N}\omega_i g(X_{1:T}^{(i)}) \leq \delta_0$ and $\frac{1}{N}\sum_{i=1}^{N}\omega_i X_{1:T}^{(i)} \in G$.

3.6 Planning in Strong Winds

The lighter than air robot could be driven with its own actuation, but due to its large inertia and slow dynamics, its actuation capability is fairly limited when winds are too strong. In some missions [32, 221], it would be more efficient to take advantage of the wind field and ride on the wind that is much stronger than what the actuator would produce. However, path planning in a flow field that is much stronger than the vehicle's actuation limits is very little studied. One of the challenges in dealing with wind field is that it is driven by many factors and therefore is highly nonlinear and time varying. It is possible to formulate the path planning as a graph search problem on a directed graph by discretizing the space time world and the vehicle actuation. The following discussion was mostly taken from [118, 221].

As the time constant of the vehicle dynamics is significantly smaller than the time constant of the path planning problem, the vehicle is assumed to move with the wind when no actuation is applied. The wind field $W(r,t)$ is highly nonlinear.

In order to handle this nonlinearity, the trajectory planning problem is converted into a graph search problem on a discretized environment. Because the wind at a given location differs depending when the lighter than air robot reaches there temporal and spatial discretizations are performed.

- Node: Let s_i denote the ith node in the graph. Each node s_i is a function of the position x_i, y_i, z_i and the time t_i. A uniform grid is used to represent the environment and let n_x, n_y, n_z, n_t respectively denote the number of cells in x, y, z and time axes. Furthermore, let Δx, Δy, Δy, Δt respectively denote the discretized step size along these axes. Define $n_r = n_x n_y n_z$ as the number of cell positions in the environment and $N = n_r n_t$ as the total number of cells $(i = 1..N)$
- Edge: If the winds were stationary, one could find which adjacent cell the lighter than air robot will hit from a given cell by looking only at the wind direction at the cell center. The edge cost in this section is a time of travel from one node to the next and could be similarly computed by simply calculating the distance to the next cell and dividing it by the wind velocity. However, this does not apply to the time varying wind case because the wind direction/magnitude could change as the lighter than air robot travels.

The idea is to numerically integrate the velocity vector over time with a time step Δt. To simplify the planning problem, the actuation is assumed to be constant over this duration. From each cell, n actuation vectors are considered, that are different in magnitude and/or direction. From each actuation vector, the integration starts from the center of each cell s_i and goes until the integrated position reaches one of the neighboring cells s_i'. The integration gives both the neighboring cell s_i' that the vehicle will go next and the time of travel from s_i to s_i'. Let $c(s_i)$ denote this time of travel, which is also used as a cost. To account for a continually weak wind field in which the lighter than air robot cannot reach a neighboring cell, the integration is terminated at time $n_{t_{\max}}\Delta t$ so that $c(s_i) < n_{t_{\max}}\Delta t$, $\forall i$. In such a case, the vehicle is assumed to stay at the same location for the first time step, reaching a node with the same position by a different time $t_i + \Delta t$.

Let $R(s_i)$ denote a set of reachable cells from s_i. Then, using the information of node s_i', $R(s_i)$ is characterized as

$$s_j \in R(s_i) \Rightarrow \begin{cases} x_j = x_i' \\ y_j = y_i' \\ z_j = z_i' \\ t_j = t_i + c'(s_i) \end{cases} \tag{3.70}$$

From each node s_i, the connections made to all nodes in $R(s_i)$ with the associated cost $c(s_i)$. Let A denote this weighted adjacency matrix of the graph. As the lighter than air robot only moves in the positive direction in time, the adjacency matrix of the graph can be represented with an upper block triangular matrix and this upper level block triangular structure can be exploited to decompose a graph search problem.

Let n_s denote the number of starting nodes. The problem statement for the graph search is to find the shortest paths from n_s starting nodes s_i to all n_r locations. Once the graph is constructed, an A* (or D*) algorithm can be applied to find the shortest paths. A matrix of size $n_s \times n_r$ can represent the minimum time of arrival at each 3D location from each starting node. Let this matrix be denoted by C^*. Its (p, q) element C_{pq} stores the time of travel from the pth starting location to the qth location and is set to be ∞ if there exists no such trajectory. Each block contains a snapshot of the 3D world whose size is $n_r \times n_r$. With a large n_t, the memory requirement for graph construction and A* (or D*) algorithm becomes significant. However, the upper block triangular structure of the adjacency matrix can be exploited to decompose the problem with several smaller subproblems that uses much smaller memory.

The main idea of the decomposition algorithm is as follows. The vehicle is assumed to start at $t = 0$, so that all the starting nodes are in the $(1, 1)$ block. This approach splits the weighted adjacency matrix A into several sub-matrices M_k, $k = 1 \dots k_{\max}$ and repeatedly applies A* (or D*) algorithm to each sub-matrix. let b_r and b_c be the number of row blocks and column blocks in each sub-matrix M_k respectively. Because each block has a size $n_r \times n_r$, the sub-matrix M_k consists of row from $(k-1)b_r n_r + 1$ to $k b_r n_r$ and columns from $(k-1)b_r n_r + 1$ to $(k-1)b_r n_r + b_c n_r$ of A. The result of each subproblem can be represented by a matrix whose size is much smaller than the sub-matrix used in the subproblems. The next subproblem is formed by appending to its sub-matrix the small matrix obtained in the previous sub-problem. This process is repeated until all the sub-matrices are processed or all the shortest path from the starting node to the 3D locations is found.

Algorithm (Decomposed Dijkstra)

1. Solve for shortest paths with weighted adjacency matrix M_1, obtain a matrix $[E_1|F_1]$ with the minimum cost.

2. From E_1, compute the minimum path cost C_1 from n_s starting locations to each 3D location.
3. Initialize the cost matrix $C^* = C_1$
4. For k = 2 to $k_{\max}$ do
5. if all elements of C^* are not ∞ then
6. break
7. end if
8. Solve for shortest paths with weighted adjacency matrix

$$\begin{bmatrix} O_{n_s,n_s}| & F_{k-1}| & O_{n_s,b_r n_r} \\ O_{b_r n_r,n_s}| & M_k & \end{bmatrix} \tag{3.71}$$

Obtain a matrix $[G_k|E_k|F_k]$ with the minimum cost.

9. From E_k compute the minimum path cost from starting locations to each 3D location C_k
10. Update the cost matrix $C^* = \min(C^*, C^k)$. Note that this min operation is element wise.
11. end for

The first step (line 1) is to run Dijkstra's algorithm with M_1 as a weighted adjacency matrix: sub-problem 1. The resultant shortest path costs (from n_s starting locations to all the cells considered in sub-problem 1) can be represented by a matrix D_1 whose size is of $n_s \times (b_c n_r)$. This D_1 can be partitioned into two matrices $D_1 = [E_1|F_1]$ where E_1 is a matrix of size $n_s \times (b_c - b_r)n_r$ and F_1 is a matrix of size $n_s \times (b_c - b_r)n_r$.

The matrix E_1 corresponds to cells with time steps between 0 and $(b_r - 1)\Delta t$ and the optimal paths to these cells have been obtained. The matrix F_1 corresponds to cells with time stamps after $b_r \Delta t$ and not all the options to reach them have been explored yet.

Using E_1 the cost matrix C^* is initialized. E_1 is then reshaped into a 3D array of size $n_s \times n_r \times b_r$ and then take the minimum along the 3rd dimension, obtaining a $n_s \times n_r$ matrix denote by C_1. For the nodes that cannot be reached from the starting cells, Dijkstra's algorithm is assumed to output ∞ as the cost. The cost matrix C* is then set to be $C^* = C_1$, in line 3. The next sub-problem uses M_2 and considers times from $b_c\Delta t$ to $(2b_c - 1)\Delta t$ but it also includes n_s starting nodes. Therefore, in order to form the weighted adjacency matrix of this sub-problem M_2 must be augmented using the result of the previous sub-problem (line 8). Let $O_{p,q}$ denote a zero matrix of size $p \times q$. Then, the weighted adjacency matrix used in sub-problem 2 is written as:

$$\begin{bmatrix} O_{n_s,n_s}| & F_1| & O_{n_s,b_r n_r} \\ O_{b_r n_r,n_s}| & M_2 & \end{bmatrix} \tag{3.72}$$

Note that an infinity matrix of width n_s is appended from the left. This is because the starting nodes at time $t = 0$ have only outgoing edges. The matrix F_1 obtained in the previous sub-problem, stores the cost of moving from the starting cells to all cells at time between $(b_r + 1)\Delta t$ and $b_c \Delta t$.

After solving sub-problem 2, the resultant shortest path costs are represented by a matrix D_2. The first n_s columns of D_2 correspond to the starting cells, so that D_2 can be partitioned into $D_2 = [G_2|E_2|F_2]$ where G_2 is a square matrix of size n_s and E_2 and F_2 respectively the same size as E_1 and F_1.

From E_2, the minimum path costs from starting locations to each 3D locations are computed and C^* is updated (lines 9–10). As shown in the algorithm, the process is repeated until all the rows of the adjacency matrix A are used or all paths to the 3D locations are found and C^* has no non ∞ element.

3.7 Task Assignment

At the level of the task assignment problem, the vehicle dynamics is usually abstracted as being first order with a maximum speed $V_{\max}$. This abstraction allows the task assignment algorithm to capture important aspects of the vehicle performance: in particular, how long the lighter than air robot is expected to take to fulfill a particular task, while being sufficiently simple to allow computational tractability. The trajectory planning and control levels below the task assignment are responsible for carrying out these lower-level functions, allowing this simplification to be made. The vehicle propulsion system may be abstracted as an entity that enables the vehicle to move at the maximum speed $V_{\max}$. Health feedback about the propulsion system may dynamically modify $V_{\max}$ to reflect the state of the propulsion system. For example, knowledge of a failing motor may cause $V_{\max}$ to decrease from its nominal value. Knowledge of the fuel state of the vehicle is important to be able to estimate the remaining useful flight time of the vehicle. The performance model should include an estimator that performs the remaining flight time calculation based on the remaining fuel, average fuel consumption rates, and perhaps other environmental factors. Use of this information allows the task assignment algorithm to safely make assignments while ensuring the lighter than air robot can return to the base before running out of fuel [30, 117].

Issues related to how vehicle health (e.g. fuel management and vehicle failures) affects the real time mission planning, are introduced. This represents a step toward enabling robust decision making for the aerial vehicle by improving its operational reliability and capability through better self-awareness and adaptive mission planning. One such solution architecture is shown in Fig. 3.1 in which a number of components are combined to achieve the overall goals of the mission. The mission planning component provides the list of tasks to the task assignment component [3]. Once the assignments have been made, they are sent to the trajectory designer which plans feasible trajectories. The output of the trajectory designer is a sequence of way-points for the lighter than air robot to follow. These way-points are sent to the controllers which compute the actual controls needed to follow the way-points plans [128].

Inherent in each of the component in the architecture is a set of interconnected models used to predict future system behavior. For example, the controller contains a model of the control input dynamics of the vehicle, while the task assignment

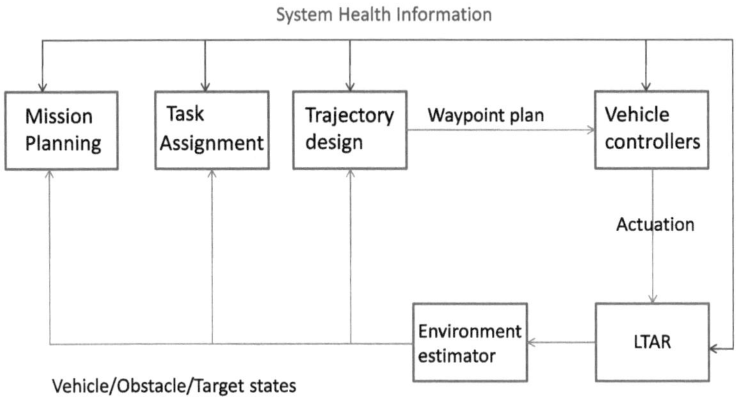

Fig. 3.1 Overall autonomous mission system architecture

component contains a model of the performance the LTAR is expected to produce if given an assigned task. In the most general sense, system actions are selected by searching for actions that lead to desirable, predicted outcomes as given by the system models. Clearly, the performance of the system, therefore, depends heavily on the accuracy of these models.

One strategy for improving the accuracy of the models is to include additional feedback loops that provide information that can be used to adjust the models in real time. The amount, type, and quality of feedback information that each component receives plays a large role in how effectively the system can deal with dynamically changing factors in the environment, mission objectives and state of the vehicles. Intuitively, feedback is necessary wherever there is uncertainty in the system, so that the initial plan of action made by each of the components of the planner can be modified when changes occur. Uncertainty may be present at all levels of the planning architecture as a result of incomplete knowledge of many factors, such as actuator performance at the control level, dynamic constraints at the trajectory design level, sensor health at the task assignment level and long-term maintenance needs at the mission management level...

Most of the work done on task assignment has used only a static vehicle performance model, making it difficult for these approaches to adapt to unexpected changes, such as sensor failures, during the course of the mission. By updating the performance model of an already existing algorithm, previous work on the task assignment problem can be extended without requiring the modification of the existing algorithm. Its performance can be improved only by improving the quality of information available to make assignments.

The Receding horizon task assignment (RHTA) algorithm works as follows: Given the set of tasks W, distances between tasks $d(i, j)$, the algorithm enumerates all possible task sequences of specified length n_c. These sequences are called petals. The value of each petal is estimated as:

$$S_{v_p} = \sum \lambda T_{ip} S_{w_d} \tag{3.73}$$

where T_{ip} is the time at which the task i is completed in petal p, S_{w_d} is the task value and λ is a time discount factor. Given the values of all the petals S_{v_p}, receding horizon task assignment solves the following optimization problem to select the optimal petal for the lighter than air robot.

$$\max J = \sum_{p=1}^{N_{vp}} \sum_{v=1}^{N_v} S_{v_p} x_{v_p}$$

subject to

$$\sum_{v=1}^{N_v} \sum_{p=1}^{N_{vp}} A_{v_p} x_{v_p} \leq 1, \quad x_{v_p} \in \{0, 1\} \tag{3.74}$$

$$\sum_{p=1}^{N_{vp}} x_{v_p} = 1, \quad \forall v \in 1, \dots, N_v$$

Here x_{v_p} is a binary variable that is equal to 1 if the pth petal is selected and 0 if not and A_{v_p} equals 1 if task i is visited in petal p and 0 otherwise.

In this example, health state information is represented by adding a fuel state to the vehicle model. In this case, the fuel model is straightforward;

- The vehicle fuel level f_i decreases at a constant rate k_{fuel} anytime the vehicle is flying
- If f_i reaches zero before the vehicle refuels, LTAR will reverse to a free balloon.
- In addition, the occurrence of failures is modeled as a Poisson process with time intensity ρ_f; when a failure occurs, the rate of fuel burn increases to $k_{fuel,failure} > k_{fuel}$. Thus this failure mode increases the rate at which fuel is burned (and thus decreases the time at which a vehicle can complete tasks).

Due to the inclusion of randomly occurring failures, the fuel model is able to capture some of the uncertainty in the health state of the vehicle.

The Receding horizon task assignment will assign the Lighter than air robot to return to the base when every possible permutation of way-points is rejected by the pruning criterion. Thus this method provides a simple rule that determines when the vehicle should return to the base for refueling since it cannot safely service any of the remaining tasks.

3.8 Conclusions

Given the flight management system goals and environment, some of the quantities may be perceived as more important than others and weighted more heavily, resulting in a behavior that is optimal for this LTAR, with these goals, in this particular environment. A computational modeling system is presented, that recognize what kind of behaviors is mostly likely to be successful, given the goal and the environment at the moment.

In a typical aerial autonomous vehicles application, the vehicle operates in 3D space and has differential constraints, including limited speed and maximum acceleration. There does not exist yet an algorithm that provides an exact analytic solution to such a problem. Even state of the art approximation algorithms operating in a 3D subspace of this problem space are difficult to compute in real time. Furthermore, several simplifications and sub-cases of the general problem have been proven to be unsolvable in polynomial time. Approximation algorithms are possible and often rely on exact solutions to simplified sub-problems. Some collision avoidance methods are presented.

Flight management of a vehicle in a dynamic partially known environment is a complex real time problem. The flight route can be preprogrammed at launch based on current weather conditions and forecast information, but once en route, changes to the flight plan have to be made in accordance to changes in the weather. Even though forecast models have improved, weather and wind patterns continuously change and are still difficult to predict. Thus, the use of frequently updated weather information during the flight will allow adaptation and re-planning based on current and near-term forecasts. The Weather forecast become an important part of the flight management system, to detect signs of a slight change in wind velocity, monitor development and movement of clouds, recognize conditions suitable for the development of temperature inversions or of local wind phenomena.

Given the uncertain nature of real-world problems, the ability to deal with various types of uncertainty and systems errors is also crucial for algorithms in practical applications. Another way to deal with uncertainty is to have adaptive algorithms which do not require re-computing. For most problems considered for aerial autonomous vehicles tasks, it has been proven that no algorithm exists that can find exact solutions in polynomial time. Ultimately, an algorithm's performance is judged by the operators of the actual vehicle in the field; this practical perspective needs to be kept in mind when analyzing computational complexity.

Despite the results presented in this chapter, significant research efforts are still needed to advance the state of the art of trajectory planning. Properly dealing with uncertainty has not yet been adequately studied for lighter than air robots planning.

Many of the algorithms designed to solve the dynamics-constrained problem rely on a decomposition approach, first solving an obstacle avoidance problem (presented in this chapter) then a path planning problem applying smoothing constraints (presented in Chap. 4), forming a trajectory that conforms to the path, and using a control loop to follow this trajectory (presented in Chap. 5).

Chapter 4
Trajectory Design

Abstract As the lighter than air robot is under-actuated, i.e. less control inputs than degrees of freedom, the set of feasible trajectory will be restricted and consequently, the problem of trajectory generation becomes more complicated than a simple interpolation. Care must be taken in the selection of the basic primitives.

4.1 Introduction

Classically, methods such as continuous optimization and discrete search are sought in trajectory generation. In general, the optimality of a trajectory can be defined according to several objectives, like minimizing the transfer time or the energy [16, 18, 29, 51, 85, 144]. Trajectories are optimized by the application of numerical optimal control methods, based on the calculus of variations. Dubins in [58] considered a particle moving at a constant velocity in the plane with a constraint of trajectory curvature. He proved the existence of shortest paths for his problem and showed that the optimal trajectories are a combination of arc of circles and segments of lines. Boukraa et al. in [38] presents a 3D trim trajectories planner algorithm for an autonomous plane. The proposed algorithm uses a sequence of five elementary trim trajectories to generate a 3D global trajectory in space. Trim trajectories take a particular place in the aeronautic applications [28, 82, 83, 89]. They correspond to forces and moments equilibrium in body-fixed frame under a fixed controls configuration. More precisely, they correspond to the equilibrium points of dynamic equations. The approach will be presented in the second section.

Frazzoli et al. in [72] described motion plans as the concatenation of a number of well defined motion primitives selected from a finite library. They use a maneuver automaton, defining rules for the concatenation of primitives in the form of a regular language, a maneuver being defined as a non trivial primitive compatible from the beginning and the end with trim primitives. In this scope, the problem of trajectories planning is subdivided into two levels. The lower level consists in the computation of the set of basic primitives, to be used by the higher level to construct the nominal trajectory by a convenient sequencing strategy. Sufficient conditions are given in [71, 82, 222] to guarantee minimal set of trim trajectories to ensure a controllability in the symmetric space.

In the papers cited above, the atmosphere was considered to be an isotropic and homogeneous medium, i.e. when there is no wind and the air density is constant with

Y. Bestaoui Sebbane, *Lighter than Air Robots*,
Intelligent Systems, Control and Automation: Science and Engineering 58,
DOI 10.1007/978-94-007-2663-5_4, © Springer Science+Business Media B.V. 2012

altitude. However, wind cannot be ignored. McGee and Hedrick in [144] describe a method for finding the minimum time path from an initial position and orientation to a final position and orientation in the 2D plane for an airplane with a bounded turning rate in the presence of a known constant wind with a magnitude less than the airplane velocity. The problem statement is equivalent to finding the minimum time path from an initial configuration to a final one, over a moving virtual target, where the velocity of the virtual target is equal and opposite to the velocity of the wind. Nelson et al. in [158] have introduced a method for a mini aerial vehicle path following based on the concept of vector field in the presence of constant wind disturbances.

Rysdyk in [180] presents a path formulation for maneuvering a fixed wing aircraft in wind. Wind refers to an un-accelerated horizontally moving air mass. The inertial path of a fixed wing aircraft circling in wind can be formulated as a trochoid curve. In these papers, only 2D horizontal motion was considered. Seube et al. in [187] formulated the take-off problem in a 2D vertical plane in the presence of wind shear as a differential game against nature. The first player is the relative angle of attack of the aircraft (considered as the control variable) and the second player is the disturbance caused by a wind shear. Zhao and Qi in [229] studied the optimal powered dynamic soaring flights of UAV that use low altitude wind gradients for reducing fuel consumption. Optimal control methods are used to study fuel-efficient dynamic soaring flights that utilize both wind energy and engine thrust.

This chapter consists of 6 sections. Section 4.2 presents the analysis of the hovering trajectories in 3D, then, in Section 4.3, lateral planning of the lighter than air robot in cruising flight followed in Section 4.4 by trajectory design for translational dynamics when wind is considered. Efforts are put in this paragraph on a constant velocity wind. Section 4.5 introduces parametric curves and explicits the relationship between trajectory generation algorithms and under-actuation. Finally, some conclusions and perspectives are the subject of Section 4.6.

4.2 Trajectory Generation in Hover

In a trimmed maneuver, the lighter than air robot will be accelerated under the action of non-zero resultant aerodynamic and gravitational forces and moments, these effects will be balanced by effects such as centrifugal and gyroscopic inertial forces and moments. Under the trim condition, the vehicle motion is uniform in the body fixed frame. The trim trajectories have the advantage of facilitating the planning and control problems. The aerodynamic coefficients which are variable in time and space become stationary under this condition and their identification becomes easier [172].

4.2.1 Trim Trajectories

The remainder of this section gives a mathematical characterization of the trim trajectories [15, 82, 83].

4.2.1.1 Trim Trajectories Characterization

As mentioned before, trim trajectories are characterized by the stationarity of the body-fixed velocity components and the controls. This condition can be mathematically formalized by:

$$\dot{V}(t) \equiv 0 \qquad \dot{\Omega}(t) \equiv 0 \quad \forall t \in [0, t_f] \tag{4.1}$$

Focusing on the angular velocity kinematics transformation:

$$\Omega = J(\eta_2)^{-1}\dot{\eta}_2 \tag{4.2}$$

By deriving the above equation with respect to time and using condition (4.1), the following relations can be written:

$$\begin{aligned}
\dot{p} &= \ddot{\phi} - \ddot{\psi}\sin\theta - \dot{\psi}\dot{\theta}\cos\theta = 0 \\
\dot{q} &= \ddot{\theta}\cos\phi - \dot{\theta}\sin\phi\dot{\phi} + \ddot{\psi}\sin\phi\cos\theta \\
&\quad + \dot{\psi}\dot{\phi}\cos\phi\cos\theta - \dot{\psi}\dot{\theta}\sin\phi\sin\theta = 0 \\
\dot{r} &= -\ddot{\theta}\sin\phi - \dot{\theta}\dot{\phi}\cos\phi + \ddot{\psi}\cos\phi\cos\theta \\
&\quad - \dot{\psi}\dot{\phi}\sin\phi\cos\theta - \dot{\psi}\dot{\theta}\cos\phi\sin\theta = 0
\end{aligned} \tag{4.3}$$

Based on the dynamics equations, all forces and moments acting on the lighter than air robot depending on the velocity vector are constant except the vector of the aerostatic forces and moments τ_s which depends on the attitude variables, the roll ϕ and the pitch θ angles. It follows that, in order to guarantee the stationarity of this vector: the roll angle ϕ_e and pitch θ_e angle must be constant. Hence, Eq. (4.3) can be simplified to:

$$\begin{aligned}
p_e &= -\dot{\psi}_e \sin\theta_e \\
q_e &= \dot{\psi}_e \sin(\phi_e)\cos(\theta_e) \\
r_e &= \dot{\psi}_e \cos(\phi_e)\cos(\theta_e)
\end{aligned} \tag{4.4}$$

The components of the body-fixed angular velocity vector depend on the roll angle ϕ_e, pitch angle θ_e and the yaw rate $\dot{\psi}_e$.

It is thus possible to characterize the geometry of the trim trajectories as follows:

$$\dot{\eta}_1 = R(\eta_2)\begin{pmatrix} u_e \\ v_e \\ w_e \end{pmatrix} = R_z(\psi)\begin{pmatrix} a \\ b \\ c \end{pmatrix} \tag{4.5}$$

where

$$\begin{pmatrix} a \\ b \\ c \end{pmatrix} = R_y(\theta_e)R_x(\phi_e)\begin{pmatrix} u_e \\ v_e \\ w_e \end{pmatrix}$$

Hence, when the translational kinematics are used, the following relation is obtained:

$$\dot{\eta}_1 = \begin{pmatrix} a\cos(\dot{\psi}_e t + \psi_1) - b\sin(\dot{\psi}_e t + \psi_1) \\ a\sin(\dot{\psi}_e t + \psi_1) + b\cos(\dot{\psi}_e t + \psi_1) \\ c \end{pmatrix} \tag{4.6}$$

where ψ_1 is an integration constant. This equation can also be written under the following form

$$\dot{\eta}_1 = \begin{pmatrix} V_e\cos(\gamma_e)\cos(\dot{\psi}_e t + \psi_0) \\ V_e\cos(\gamma_e)\sin(\dot{\psi}_e t + \psi_0) \\ -V_e\sin(\gamma_e) \end{pmatrix} \tag{4.7}$$

$\gamma_e = \text{acos}(\frac{\sqrt{a^2+b^2}}{V_e})$ represents the flight path angle. Or equivalently

$$\sin\gamma_e = \frac{-\cos\phi_e\cos\theta_e u_e + \sin\phi_e v_e + \cos\phi_e\sin\theta_e w_e}{\sqrt{u_e^2 + v_e^2 + w_e^2}}$$

The navigation velocity is given $V_e = \|V\|$, and ψ_0 is the initial heading angle.

Integrating the above equation, the geometric characterization of trim trajectories can be described by:

$$\eta_1 = \begin{pmatrix} \frac{V_e}{\dot{\psi}_e}\cos(\gamma_e)\sin(\dot{\psi}_e t + \psi_0) \\ -\frac{V_e}{\dot{\psi}_e}\cos(\gamma_e)\cos(\dot{\psi}_e t + \psi_0) \\ -V_e\sin(\gamma_e)t \end{pmatrix} + \begin{pmatrix} x_1 \\ y_1 \\ z_1 \end{pmatrix} \tag{4.8}$$

where the integration constants are given by

$$\begin{aligned} x_1 &= x_0 - \frac{V_e}{\dot{\psi}_e}\cos\gamma_e\sin\psi_e \\ y_1 &= y_0 + \frac{V_e}{\dot{\psi}_e}\cos\gamma_e\cos\psi_e \\ z_1 &= z_0 \end{aligned}$$

where $(x_0, y_0, z_0)^T$ is the initial position of the lighter than air robot. It is possible to parametrize the trim trajectory by the vector $\mathcal{T}_e = (\phi_e, \theta_e, u_e, v_e, w_e, \dot{\psi}_e)$ where the subscript e denotes the equilibrium values.

This trajectory can also be expressed with respect to the curvilinear abscissa s as for a uniform motion: $s = V_e t$

$$\eta_1 = \begin{pmatrix} \frac{V_e}{\dot{\psi}_e}\cos(\gamma_e)\sin(\frac{\dot{\psi}_e}{V_e}s + \psi_0) \\ -\frac{V_e}{\dot{\psi}_e}\cos(\gamma_e)\cos(\frac{\dot{\psi}_e}{V_e}s + \psi_0) \\ -\sin(\gamma)s \end{pmatrix} + \begin{pmatrix} x_1 \\ y_1 \\ z_1 \end{pmatrix} \tag{4.9}$$

Depending on the values of γ_e and $\dot{\psi}_e$, the trajectories can be represented by a helix (see Fig. 4.3), a circle arc (see Fig. 4.2) or a straight line (see Fig. 4.1).

4.2.1.2 Trim Trajectories Calculation

The above kinematic analysis of the trim trajectories shows that their geometry depends on the body-fixed linear velocity vector V_e, the roll angle ϕ_e, pitch angle θ_e and the rate of yaw angle $\dot{\psi}_e$. The choice of these quantities should satisfy the dynamic equations, the controls saturations and envelope protection constraints.

Control Saturation Constraints The lighter than air robot controls belong to specific ranges due to the power, angle of vectorization and deflection angles of control surfaces limitations. Assuming that the option of tail rotor is not offered for this particular vehicle, the limitation constraints imposed on the force F_{m_e} and the angle of vectorization μ_{m_e} are expressed in terms of trim vector components $\mathcal{T}_e$ and control surfaces angles of deflection as follows:

$$\tau_p = \begin{pmatrix} F_{m_e}\cos(\mu_{m_e}) \\ 0 \\ -F_{m_e}\sin(\mu_{m_e}) \\ 0 \\ F_{m_e}\cos(\mu_{m_e})O_z + F_{m_e}\sin(\mu_{m_e})O_x \\ 0 \end{pmatrix} = \begin{pmatrix} F_x(\nu_e, \eta_{2_e}, \delta_e) \\ F_y(\nu_e, \eta_{2_e}, \delta_e) \\ F_z(\nu_e, \eta_{2_e}, \delta_e) \\ M_x(\nu_e, \eta_{2_e}, \delta_e) \\ M_y(\nu_e, \eta_{2_e}, \delta_e) \\ M_z(\nu_e, \eta_{2_e}, \delta_e) \end{pmatrix} \tag{4.10}$$

$$F_{m_{min}} \le F_{m_e} = \sqrt{F_x^2(\nu_e, \eta_{2_e}, \delta_e) + F_z^2(\nu_e, \eta_{2_e}, \delta_e)} \le F_{m_{max}} \tag{4.11}$$

$$\mu_{min} \le \mu_{m_e} = \operatorname{atan} \frac{F_z(\nu_e, \eta_{2_e}, \delta_e)}{F_x(\nu_e, \eta_{2_e}, \delta_e)} \le \mu_{max} \tag{4.12}$$

and the limitations of control surfaces angles of deflection are given by:

$$\delta_{e_{min}} \le \delta_{e_e} \le \delta_{e_{max}} \tag{4.13}$$

$$\delta_{r_{min}} \le \delta_{r_e} \le \delta_{r_{max}} \tag{4.14}$$

Under-actuation Constraints The control of the lighter than air robot is performed by 4 actuators (i.e. throttle, vectored angle, rudders and elevators). Hence, in this case, the under-actuation degree is 2. The action of rudders and elevators on the dynamic equation appears in a nonlinear manner, making the derivation of the under-actuated constraints very difficult. To bypass this problem, the control surfaces (rudders and elevators) angles are considered in the optimization variables.

The under-actuated constraints related to propulsion system is given by:

$$\begin{pmatrix} M_y(\nu_e, \eta_{2_e}, \delta_e) - O_z F_x(\nu_e, \eta_{2_e}, \delta_e) + O_x F_z(\nu_e, \eta_{2_e}, \delta_e) \\ F_y(\nu_e, \eta_{2_e}, \delta_e) \\ M_x(\nu_e, \eta_{2_e}, \delta_e) \\ M_z(\nu_e, \eta_{2_e}, \delta_e) \end{pmatrix} = 0 \tag{4.15}$$

Other Constraints The last type of constraints considered in the calculation of trim trajectory concerns the geometry of the path and the speed of the lighter than air robot V_e. The geometric parameters are the flight path angle γ_e and the turning radius R_e, set to desired values by the user [71, 82]. It is possible to consider also some constraints on the angle of attack α_e, and side slip β_e angles to guarantee the flight envelope protection. These constraints are given by:

$$R_e = \frac{V_e}{|\dot{\psi}_e|} \cos(\gamma_e) \tag{4.16}$$

$$\gamma_e = \operatorname{acos}\left(\frac{\sqrt{a^2 + b^2}}{V_e}\right) \tag{4.17}$$

$$V_e = \sqrt{u_e^2 + v_e^2 + w_e^2} \tag{4.18}$$

$$\alpha_{min} \leq \alpha_e = \operatorname{asin}\left(\frac{w_e}{\sqrt{u_e^2 + w_e^2}}\right) \leq \alpha_{max} \tag{4.19}$$

$$\beta_{min} \leq \beta_e = \operatorname{asin}\left(\frac{v_e}{\|\nu_{1_e}\|}\right) \leq \beta_{max} \tag{4.20}$$

Taking into account these constraints, the calculation of the trim conditions are determined by solving a set of nonlinear equalities and inequalities for the vector $\mathcal{T}_e$ and control surfaces vector $\delta = [\delta_e, \delta_r]^T$, that makes the body-fixed accelerations vector null, i.e. $\dot{V} \equiv 0$ and $\dot{\Omega} \equiv 0$. A convenient way to do this, is to formulate an optimization problem to choose the solution that minimizes the energy consumed during the execution of the trim trajectory. This choice is straightforward because the velocity and controls are constant, and hence the energy is proportional to the arrival time. The expression of the energy is given by:

$$E = \int_{t_0}^{t_f} F_m^2 dt \tag{4.21}$$

The controls are constrained to take fixed values. It follows that:

$$E = F_m^2 (t_f - t_0) \tag{4.22}$$

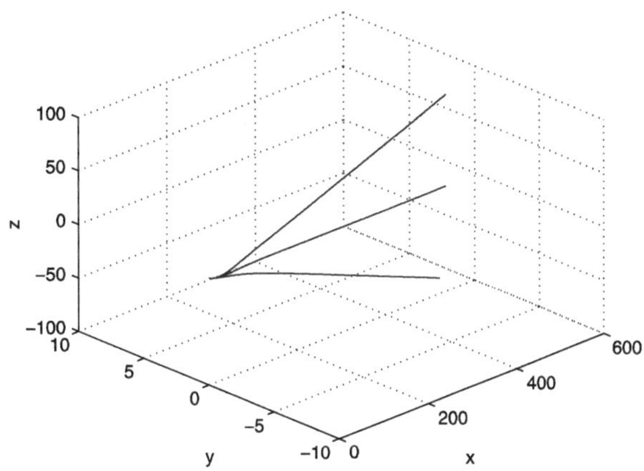

Fig. 4.1 3D trim trajectory for straight forward trim flight

4.2.1.3 Simulation Results

In this section, some simulation results obtained by the proposed algorithm are presented. The controls are constrained to $0 \leq F_m \leq 20N$, $-45° \leq \delta_e \leq 45°$, $-45° \leq \delta_r \leq 45°$ and the navigation velocity is set to $V_e = 6$ ms^{-1}. Initially the lighter than air robot is at rest. When the controls associated to required trim conditions are applied, the states of the lighter than air robot converge to the nominal values.

Straight Forward Trim Flight In this section the simulation results corresponding to straight forward trim flight situation are presented. In this case the radius of gyration is set to $R_e = \infty$. Figure 4.1 depicts the 3D trim trajectories corresponding to flight path angles $\gamma_e = -10°, 0, 10°$.

Circular Trim Trajectories with Constant Altitude The simulation results corresponding to the circular trim trajectories with constant altitude, i.e. $\gamma_e = 0$, are presented in this section. The 3D circular trim trajectories with a radius of gyration $R = 40$ m are depicted in Fig. 4.2 (red path). Due to the symmetry of the dynamics of the lighter than air robot with respect to the yaw angle ψ, it is possible to find the opposite circular trim trajectories by inverting the rudder angle δ_r (blue path).

Helicoidal Trim Trajectories In this section we present the more general case. It corresponds to the helical trim trajectories Fig. 4.3 corresponding to $R_e = 40$ m, $\gamma_e = 10°$ (blue) and $\gamma_e = -10°$ (red).

4.2.2 Under-actuation at Hover

In this paragraph, nominal trajectories are characterized, the model is supposed perfect and any perturbation such as wind or sensor disturbance is neglected. At low velocity, the surface control are ineffective so aerodynamical forces can be neglected. The three equality constraints deriving from the under-actuation are sought.

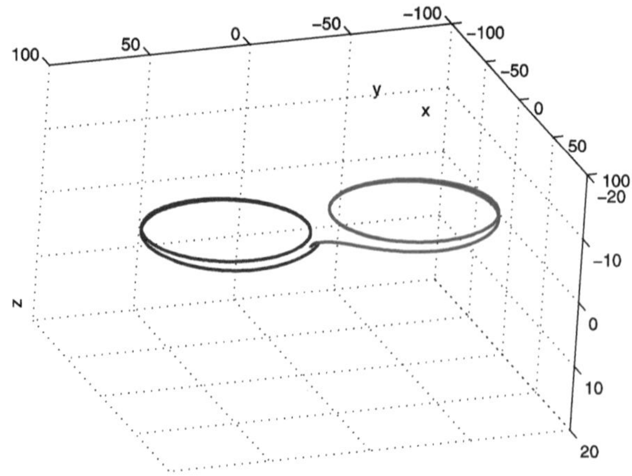

Fig. 4.2 (Color online) 3D circular trim trajectories with constant altitude

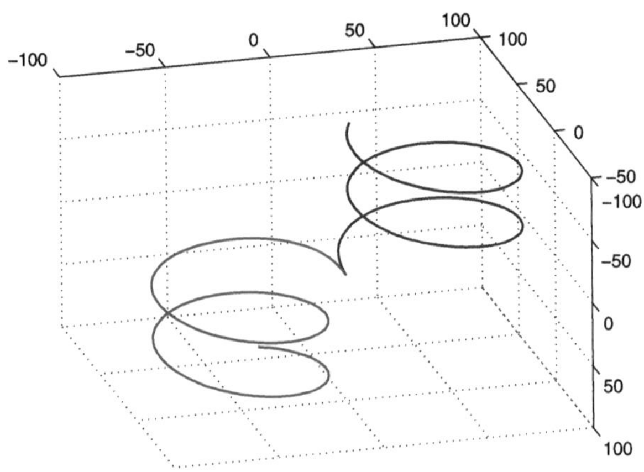

Fig. 4.3 (Color online) 3D helical trim trajectories

Remark 4.1 The following notation is used in the sequel of this section $S_\theta = \sin\theta$, $C_\theta = \cos\theta$.

4.2.2.1 Problem Formulation

Considering the dynamics of the lighter than air robot and its propulsion at low speed, the following dynamics equations can be written:

$$f = M\dot{\nu}_1 + \nu_2 \times M\nu_1 + b = \begin{pmatrix} F_M \sin\mu \\ F_T \\ F_M \cos(\mu) \end{pmatrix} \tag{4.23}$$

and

$$\tau = I\dot{\nu}_2 + \nu_2 \times I\nu_2 + \nu_1 \times M\nu_1 + \beta_1 = \begin{pmatrix} 0 \\ (O_z \cos\mu + O_x \sin\mu)\, F_M \\ O_y F_T \end{pmatrix} \tag{4.24}$$

with

$$b = \mathrm{diag}(D_v)\nu_1 + (mg - B)\begin{pmatrix} C_\psi C_\theta \\ S_\psi C_\theta \\ -S_\theta \end{pmatrix}$$

$$\beta_1 = \mathrm{diag}(D_\Omega)\nu_2 + B\begin{pmatrix} C_\psi C_\theta \\ S_\psi C_\theta \\ -S_\theta \end{pmatrix}$$

and

$$\mathrm{Diag}(D_v) = -\,\mathrm{diag}(X_u, Y_v, Z_w)$$
$$\mathrm{Diag}(D_\Omega) = -\,\mathrm{diag}(L_p, M_q, N_r)$$

The roll moment being zero (see Eq. (4.23)), $\tau_1 = 0$ gives the following constraint:

$$I_{xx}\aleph + (I_{xx} - I_{zz})\left(-\dot{\theta}C_\phi + \dot{\psi}S_\phi C_\theta\right) \left(-\dot{\theta}S_\phi + \dot{\psi}C_\phi C_\theta\right) + D_p\left(\dot{\phi} - \dot{\psi}S_\theta\right) + z_b B C_\theta S_\phi = 0 \tag{4.25}$$

where

$$\aleph = \ddot{\phi} - \ddot{\psi}S_\theta - \dot{\psi}\dot{\theta}C_\theta \tag{4.26}$$

In addition, $(O_z \cos\mu + O_x \sin\mu)F_M = \tau_2$ gives

$$\begin{aligned} I_{yy}\aleph' &+ (M_{11} - M_{33})uw + (I_{xx} - I_{zz})\left(\dot{\phi} - \dot{\psi}S_\theta\right)\left(-\dot{\theta}S_\phi + \dot{\psi}C_\phi C_\theta\right) \\ &+ D_q\left(\dot{\theta}C_\phi + \dot{\psi}S_\phi C_\theta\right) + z_b B S_\theta + O_z M_{11}\dot{u} \\ &+ O_z M_{22} w\left(\dot{\theta}C_\phi + \dot{\psi}S_\phi C_\theta\right) - O_z M_{22} v\left(-\dot{\theta}S_\phi + \dot{\psi}C_\phi C_\theta\right) \\ &+ O_z D_u u + O_z(mg - B)S_\theta = 0 \end{aligned} \tag{4.27}$$

where

$$\aleph' = \ddot{\theta}C_\phi - \dot{\theta}\dot{\phi}S_\phi + \ddot{\psi}S_\psi C_\theta + \dot{\psi}\dot{\phi}C_\phi C_\theta - \dot{\psi}\dot{\theta}S_\phi S_\theta \tag{4.28}$$

The third constraint can be derived from the constraint: $O_y F_T = \tau_3$ and can expressed as:

$$\begin{aligned} I_{zz}\aleph'' &+ (M_{22} - M_{11})uv + (I_{yy} - I_{xx})\left(\dot{\phi} - \dot{\psi}S_\theta\right)\left(\dot{\theta}C_\phi + \dot{\psi}S_\phi C_\theta\right) \\ &+ N_r\left(-\dot{\theta}S_\phi + \dot{\psi}C_\phi C_\theta\right) \\ &+ O_y M_{22}\dot{v} - O_y M_{22} w\left(\dot{\phi} - \dot{\psi}S_\theta\right) \\ &+ O_y M_{11} u\left(-\dot{\theta}S_\phi + \dot{\psi}C_\phi C_\theta\right) + O_y D_v v - O_y(mg - B)S_\phi C_\theta = 0 \end{aligned} \tag{4.29}$$

where,

$$\aleph'' = -\ddot{\theta}S_\phi - \dot{\theta}\dot{\phi}C_\phi + \ddot{\psi}C_\phi C_\theta - \dot{\psi}\dot{\phi}S_\phi C_\theta - \dot{\psi}\dot{\theta}C_\phi S_\theta \tag{4.30}$$

4.2.2.2 Analysis of the Constraints

The following formulation of the differential equations is considered. The variations of the roll and pitch angles as well as the longitudinal velocity are imposed and the influence of the under-actuation on the variations of the yaw angle, the lateral and vertical velocities are studied.

The first equality constraint (see Eq. (4.25)) is equivalent to the resolution of an ordinary differential equation of the from:

$$a(t)\ddot{\psi} + b(t)\dot{\psi}^2 + c(t)\dot{\psi} + d(t) = 0 \tag{4.31}$$

where

$$\begin{aligned} a(t) &= I_{xx} S_\theta \\ b(t) &= (I_{zz} - I_{yy})(C_\phi S_\phi C_\theta^2) \\ c(t) &= -S_\theta D_p + 2(I_{zz} - I_{yy}) C_\theta C_\phi^2 - (I_{xx} + I_{zz} - I_{yy})\dot{\theta} C_\theta \\ d(t) &= -z_b B C_\theta S_\theta - I_{xx}\ddot{\phi} - D_p\dot{\phi} + (I_{zz} - I_{yy})\dot{\theta}^2 S_\phi C_\phi \end{aligned} \tag{4.32}$$

if $\Xi(t) = \dot{\psi}(t)$ then the non autonomous generalized logistic equation must be solved:

$$a(t)\dot{\Xi}(t) + b(t)\Xi(t)^2 + c(t)\Xi(t) + d(t) = 0 \tag{4.33}$$

The third equality constraint can be written as:

$$w(t) = \alpha_0 + \alpha_1 u + \alpha_2 v + \alpha_3 uv + \alpha_4 \dot{v} \tag{4.34}$$

where

$$\begin{aligned} \alpha_0 = &\frac{\alpha_0'(I_{yy} - I_{xx}) - O_x(mg - B)C_\theta S_\phi + D_r(\dot{\psi} C_\theta C_\phi - \dot{\theta} S_\phi)}{-O_y(\dot{\psi} S_\theta - \dot{\phi})} \\ &+ \frac{I_{zz}(\ddot{\psi} C_\theta C_\phi - \ddot{\theta} S_\phi - \dot{\psi}\dot{\theta} S_\theta C_\phi - \dot{\theta}\dot{\phi} C_\phi)}{-O_y M_{22}(\dot{\psi} S_\theta - \dot{\phi})} \end{aligned} \tag{4.35}$$

$$\begin{aligned} \alpha_1 &= -\frac{(\dot{\psi} C_\theta C_\phi - \dot{\theta} S_\phi) M_{11}}{(\dot{\psi} S_\theta - \dot{\phi}) M_{22}} \\ \alpha_2 &= -\frac{D_v}{(\dot{\psi} S_\theta - \dot{\phi}) M_{22}} \\ \alpha_3 &= \frac{M_{11} - M_{22}}{O_y M_{22}(\dot{\psi} S_\theta - \dot{\phi})} \\ \alpha_4 &= -\frac{1}{(\dot{\psi} S_\theta - \dot{\phi}) M_{22}} \end{aligned} \tag{4.36}$$

and

$$\alpha_0' = \left(-\dot{\psi}^2 S_\theta C_\theta S_\phi - \dot{\psi}\dot{\theta} S_\theta C_\phi + \dot{\psi}\dot{\phi} C_\theta S_\phi + \dot{\theta}\dot{\phi} C_\phi\right) \tag{4.37}$$

The second equality constraint gives

$$\beta_0 + \beta_1 u + \beta_2 u^2 + \beta_3 uv + \beta_4 v^2 + \beta_5 u\dot{v} + \beta_6 v + \beta_7 \dot{v} + \beta_8 \dot{u} = 0 \tag{4.38}$$

where

$$\begin{aligned}
\beta_0 &= \beta_0' + \beta_3'\alpha_0 \\
\beta_1 &= \beta_1' + \alpha_1\beta_3' + \alpha_0\beta_5' \\
\beta_2 &= \beta_5'\alpha_1 \\
\beta_3 &= \beta_5'\alpha_2 + \alpha_3\beta_3' \\
\beta_4 &= \beta_5'\alpha_3 \\
\beta_5 &= \beta_5'\alpha_4 \\
\beta_6 &= \beta_2' + \alpha_2\beta_3' \\
\beta_7 &= \alpha_4\beta_3' \\
\beta_8 &= \beta_4'
\end{aligned} \tag{4.39}$$

with the following relations

$$\begin{aligned}
\beta_0' &= I_{yy}\beta_0'' + (I_{xx} - I_{zz})\beta_0''' + D_q\left(\dot{\psi} C_\theta S_\phi + \dot{\theta} C_\phi\right) - B z_b S_\theta \\
&\quad + O_z(mg - B)S_\theta \\
\beta_0'' &= \ddot{\psi} C_\theta S_\phi + \ddot{\theta} C_\theta + \dot{\psi}\dot{\phi} C_\theta C_\phi - \dot{\psi}\dot{\theta} S_\theta S_\phi - \dot{\theta}\dot{\phi} S_\phi \\
\beta_0''' &= -\dot{\theta}\dot{\phi} S_\phi + \dot{\psi}\dot{\phi} C_\phi C_\theta + \dot{\psi}\dot{\theta} S_\theta S_\phi - \dot{\psi}^2 C_\theta S_\theta C_\phi \\
\beta_1' &= O_z X_u \\
\beta_2' &= -O_z M_{22}\left(\dot{\psi} C_\theta C_\phi - \dot{\theta} S_\phi\right) \\
\beta_3' &= O_z M_{22}\left(\dot{\psi} C_\theta S_\phi + \dot{\theta} C_\phi\right) \\
\beta_4' &= O_z M_{11} \\
\beta_5' &= (M_{11} - M_{22})
\end{aligned} \tag{4.40}$$

4.2.2.3 Resolution of the Differential Equations

The differential equation

$$a\ddot{\psi} + b\dot{\psi}^2 + c\dot{\psi} + d = 0 \tag{4.41}$$

where

$$\begin{aligned}
a &= -I_{xx} S_\theta \\
b &= (I_{zz} - I_{yy}) S_\phi C_\theta C_\phi C_\theta \\
c &= -D_p S_\theta \\
d &= -z_b B C_\theta S_\phi
\end{aligned} \tag{4.42}$$

admits an analytic general solution.

By using the method of separation of variables and integration by partial fractions, in the constant coefficient case, logistic equation can be solved and the behavior of all solutions is analyzed [16]

For $\phi = 0$; $\psi(t) = \psi_0 e^{-L_p t/I_{xx}}$

For $\theta = 0$; $\psi(t) = t\sqrt{\frac{Bz_b}{C\phi(I_{zz}-I_{yy})}} + \psi_0$

For the particular case, where $\dot{\psi}$ is constant, classical trim trajectories are encountered

1. $\dot{\psi}$ is constant

$$\phi = \text{constant} = \phi_0 \qquad \theta = \text{constant} = \theta_0 \qquad \psi = \psi_0 t \tag{4.43}$$

The first equality constraint becomes a second order polynomial equation:

$$b\dot{\psi}^2 + c\dot{\psi} + d = 0 \tag{4.44}$$

The third equality constraint gives

$$w(t) = \alpha_0 + \alpha_1 u + \alpha_2 v + \alpha_3 uv + \alpha_4 \dot{v} \tag{4.45}$$

where

$$\begin{aligned}
\alpha_0 &= \frac{-\dot{\psi}^2 S_\theta C_\theta S_\phi (I_{yy} - I_{xx}) + D_r \dot{\psi} C_\theta C_\phi}{O_y \dot{\psi} S_\theta} \\
&\quad + \frac{-M_{22} O_y (mg - B) C_\theta S_\phi + I_{zz} \ddot{\psi} C_\theta C_\phi}{O_y \dot{\psi} S_\theta M_{22}} \\
\alpha_1 &= -\frac{C_\theta C_\phi M_{11}}{S_\theta M_{22}} \\
\alpha_2 &= \frac{-D_v}{M_{22} \dot{\psi} S_\theta} \\
\alpha_3 &= \frac{M_{11} - M_{22}}{O_y M_{22} \dot{\psi} S_\theta} \\
\alpha_4 &= \frac{-1}{\dot{\psi} S_\theta M_{22}}
\end{aligned} \tag{4.46}$$

while the second equality constraint gives for $\dot{u} = \dot{v} = 0$,

$$v = -\frac{\beta_0 + \beta_1 u + \beta_2 u^2}{\beta_6 + \beta_3 u + \beta_4 u^2} \tag{4.47}$$

Otherwise,

$$\beta_0 + \beta_1 u + \beta_2 u^2 + \beta_3 uv + \beta_4 u^2 v + \beta_5 u\dot{v} + \beta_6 v + \beta_7 \dot{v} + \beta_8 \dot{u} = 0 \tag{4.48}$$

When $\dot{u} = \dot{v} = \dot{w} = 0$, the classical kinematics equations of the trim equations are retrieved.

2. $\dot{\psi}$ is not constant:

   In this paragraph, the roll and pitch angles are assumed to have linear variations:

$$\theta = \dot{\theta}_0 t + \theta_0 \qquad \phi = \dot{\phi}_0 t + \phi_0 \tag{4.49}$$

When the coefficients of the non autonomous logistic equation are no longer constant, no explicit solutions can be found in general and the equilibrium point may become unstable. For a study to be complete, the existence of stable periodic or stable bounded solutions is an essential part of qualitative theory of this differential equation, in order to determine non trivial solutions and study their behavior. The logistic equation with positive non autonomous bounded coefficients has exactly one bounded solution that is positive and does not tend to the zero solution.

Solving the first equality constraint, the roll moment being null, $\forall t$, implies

$$L_p \dot{\phi}_0 = 0 \quad \Longrightarrow \quad \dot{\phi}_0 = 0 \tag{4.50}$$

Rearranging the first equality constraint with this requirement gives:

$$\dot{\theta} C_{\phi_0} S_{\phi_0} = 0 \tag{4.51}$$

three cases are possible:

$$\dot{\theta}_0 = 0 \quad \text{or} \quad \phi_0 = 0 \quad \text{or} \quad \phi_0 = \frac{\pi}{2} \tag{4.52}$$

a. The first case: trim trajectories has already been studied.
b. If the roll angle is null, the following differential equations must be solved

$$\ddot{\psi} + \dot{\psi}\left(a + b\dot{\theta}_0 \frac{C_\theta}{S_\theta}\right) = 0 \tag{4.53}$$

where

$$\begin{aligned} a &= L_p / I_{xx} \\ b &= -(I_{xx} - I_{yy} - I_{zz}) / I_{xx} \end{aligned} \tag{4.54}$$

the following derivatives $\dot{\psi}$ is obtained

$$\dot{\psi}(t) = -\frac{\dot{\theta}_0 S_{\theta_0} \theta_0^b S_\theta^{-b} e^{-a\theta/\dot{\theta}_0}}{-\cosh(a\theta_0/\dot{\theta}_0) + \sinh(a\theta_0/\dot{\theta}_0)} \tag{4.55}$$

The third case $\phi_0 = \frac{\pi}{2}$ gives the following differential equations

$$\ddot{\psi} + \dot{\psi}\left(a_1 + a_2\dot{\theta}_0 \frac{C_\theta}{S_\theta}\right) + a_3 \frac{C_\theta}{S_\theta} = 0 \tag{4.56}$$

where

$$a_1 = \frac{L_p}{I_{xx}}$$
$$a_2 = \frac{I_{xx} - I_{yy} + I_{zz}}{I_{xx}} \tag{4.57}$$
$$a_3 = \frac{B z_b}{I_{xx}}$$

c. The third equality constraint gives

$$w = \delta_1 + \delta_2 uv + \delta_3 u + \delta_4 v + \delta_5 \dot{u} \tag{4.58}$$

where

$$\delta_1 = \dot{\theta}_0 \frac{I_{yy} + I_{zz} - I_{xx}}{O_y M_{22}} - \frac{N_r C_\theta}{O_y S_\theta M_{22}} - \frac{I_{zz} \ddot{\psi} C_\theta}{O_y \dot{\psi} S_\theta M_{22}}$$
$$\delta_2 = -\frac{M_{22} - M_{11}}{O_y \dot{\psi} S_\theta M_{22}}$$
$$\delta_3 = -\frac{C_\theta M_{22}}{S_\theta M_{22}} \tag{4.59}$$
$$\delta_4 = -\frac{Y_v}{\dot{\psi} S_\theta M_{22}}$$
$$\delta_5 = -\frac{1}{\dot{\psi} S_\theta}$$

Once the yaw angle is calculated, the linear and angular velocities are determined as well as the 3D path.

4.3 Lateral Planning in Cruising Flight

The lighter than air robot model consists of 12 states that can be divided into two main modes: longitudinal and lateral. The focus of this section is on lateral mode in constant altitude [82, 84]. The vectored thrusters and elevators are associated to the longitudinal navigation controller to hold the altitude and relative velocity navigation constant. The rudders allow the airship to navigate in the horizontal plan.

4.3.1 Lateral Dynamics of the Lighter than Air Robot

The model of the trimmed lateral dynamics in the body fixed frame is given by:

$$\begin{aligned} & m_y\dot{v} - (ma_z + \dot{Y}_{\dot{p}})\dot{p} - (ma_x + \dot{Y}_{\dot{r}})\dot{r} \\ & \quad = m_z w_e p - m_x u_e r + Y_e + \dot{Y}_v v + \dot{Y}_p p + \dot{Y}_r r + \dot{Y}_\delta \delta_r \\ & \qquad + (F_G - B)\phi \cos(\theta_e) \end{aligned} \tag{4.60}$$

$$\begin{aligned} & J_z\dot{r} - J_{xz}\dot{p} + (ma_x + \dot{N}_{\dot{v}})\dot{v} \\ & \quad = ma_x w_e p - ma_x u_e r + N_e + \dot{N}_v v + \dot{N}_p p + \dot{N}_r r + \dot{N}_\delta \delta_r \\ & \qquad + a_x F_G \phi \cos(\theta_e) \end{aligned} \tag{4.61}$$

$$\begin{aligned} & J_x\dot{p} - J_{xz}\dot{r} - (ma_z + \dot{L}_{\dot{v}})\dot{v} \\ & \quad = -ma_z w_e p + ma_x u_e r + L_e + \dot{L}_v v + \dot{L}_p p + \dot{L}_r r \\ & \qquad - a_x F_G \phi \cos(\theta_e) \end{aligned} \tag{4.62}$$

these equations correspond to the Lateral, Yaw and roll dynamics. V_e and v are the axial and lateral velocity components in local frame.

$$\begin{aligned} u_e &= V_e \cos\beta \approx V_r \\ v &= V_e \sin\beta \approx V_r\beta \end{aligned} \tag{4.63}$$

where $\dot{Y}_{\dot{p}}$, $\dot{Y}_{\dot{r}}$, $\dot{Y}_p$, $\dot{Y}_r$, $\dot{Y}_\delta$, $\dot{N}_{\dot{p}}$, $\dot{N}_{\dot{r}}$, $\dot{N}_p$, $\dot{N}_r$, $\dot{N}_\delta$, $\dot{N}_x$, $\dot{L}_{\dot{v}}$, $\dot{L}_v$, $\dot{L}_p$, $\dot{L}_r$ are the aerodynamic coefficients. m_i is the apparent mass in the ith direction. J_i are inertia matrix elements. m is the airship mass. θ_e is the equilibrium pitch angle. a_i are the coordinates of the center of mass in the local frame. β is the skid angle between the relative velocity V_r and x_m axis into (x_m, y_m) plane.

In the absence of wind, this angle appears when the lighter than air robot follows a path with a curvature different from zero. For a fixed rudder deflection, i.e. corresponding to the circle path, this angle takes a constant value when an equilibrium between aerodynamic moment, caused by the airship body (hull, the vertical fins and control surfaces) motion with respect to the surrounding air and the centrifugal one is established. This angle takes on small values.

In general, in lateral movement, the lighter than air robot moves with a low speed. The equilibrium between the centrifugal moment around x_m axis caused by the rudder deflection and gravitational moment is the cause of a small roll angle and rate which can be neglected. Taking these considerations into account, the model can be simplified as:

$$m_y\dot{v} - (ma_x + \dot{Y}_{\dot{r}})\dot{r} = -m_x u_e r + \dot{Y}_v v + \dot{Y}_r r + \dot{Y}_\delta \delta_r \tag{4.64}$$

$$J_z\dot{r} + (ma_x + \dot{N}_{\dot{v}})\dot{v} = -ma_x u_e r + \dot{N}_v v + \dot{N}_r r + \dot{N}_\delta \delta_r \tag{4.65}$$

and the kinematic equations are given by:

$$\begin{aligned}\dot{\psi} &= r \\ \dot{x} &= V_e \cos(\psi + \beta) \\ \dot{y} &= V_e \sin(\psi + \beta)\end{aligned} \tag{4.66}$$

The model of the lateral dynamics of the airship can be thus written as:

$$\begin{aligned}\dot{\beta} &= a_{11}\beta + a_{12}r + b_1\delta_r \\ \dot{r} &= a_{21}\beta + a_{22}r + b_2\delta_r \\ \dot{\psi} &= r \\ \dot{x} &= V_e \cos(\psi + \beta) \\ \dot{y} &= V_e \sin(\psi + \beta)\end{aligned} \tag{4.67}$$

where:

$$a_{11} = \frac{J_z \dot{Y}_v + (ma_x + \dot{Y}_{\dot{r}})\dot{N}_v}{m_y J_z + (ma_x + \dot{Y}_{\dot{r}})(ma_x + \dot{N}_{\dot{v}})}$$

$$a_{12} = \frac{1}{V_e} \frac{J_z(\dot{Y}_r - m_x u_e) + (ma_x + \dot{Y}_{\dot{r}})(\dot{N}_r - ma_x u_e)}{m_y J_z + (ma_x + \dot{Y}_{\dot{r}})(ma_x + \dot{N}_{\dot{v}})}$$

$$a_{21} = V_e \frac{-(ma_x + \dot{N}_{\dot{v}}) + \dot{N}_v m_y}{m_y J_z + (ma_x + \dot{Y}_{\dot{r}})(ma_x + \dot{N}_{\dot{v}})}$$

$$a_{22} = \frac{-(ma_x + \dot{N}_{\dot{v}})\left(\dot{N}_r - ma_x u_e\right)}{m_y J_z + (ma_x + \dot{Y}_{\dot{r}})(ma_x + \dot{N}_{\dot{v}})}$$

$$b_1 = \frac{J_z \dot{Y}_\delta + (ma_x + \dot{Y}_{\dot{r}})\dot{N}_\delta}{m_y J_z + (ma_x + \dot{Y}_{\dot{r}})(ma_x + \dot{N}_{\dot{v}})} \frac{1}{V_e}$$

$$b_2 = \frac{-(ma_x + \dot{N}_{\dot{v}})\dot{Y}_\delta + m_y \dot{N}_\delta}{m_y J_z + (ma_x + \dot{Y}_{\dot{r}})(ma_x + \dot{N}_{\dot{v}})}$$

The lateral dynamics of the airship have thus an affine structure. In the compact form the dynamics can be given by:

$$\dot{\mathbf{X}} = f(\mathbf{X}) + g\delta_r \tag{4.68}$$

where the state variable is defined as $\mathbf{X} = (\beta, r, \psi, x, y)^T$ and

$$f(\mathbf{X}) = \begin{pmatrix} a_{11}\beta + a_{12}r \\ a_{21}\beta + a_{22}r \\ r \\ V_e \cos(\psi + \beta) \\ V_e \sin(\psi + \beta) \end{pmatrix} \quad \text{and} \quad g = \begin{pmatrix} b_1 \\ b_2 \\ 0 \\ 0 \\ 0 \end{pmatrix} \tag{4.69}$$

This system is an under actuated one i.e. states are steered by a single input control. The nonholonomic constraint can be formulated as:

$$\dot{x} \sin(\beta + \psi) - \dot{y} \cos(\beta + \psi) = 0 \tag{4.70}$$

this kind of constraints are called Pfaffian Constraints.

4.3.2 Time Optimal Extremals

In this paragraph, let's introduce reference time-optimal paths for the system under study, taking into account the system dynamics and actuator capabilities. Hence, this problem can be formulated as follows:

$$\int_0^T dt = T \tag{4.71}$$

Subject to

$$\begin{gathered} \dot{\mathbf{X}} = f(\mathbf{X}) + gU \\ \mathbf{X}(0) = \mathbf{X}_0 \quad \text{and} \quad \mathbf{X}(T) = \mathbf{X}_f \\ U_{min} \leq U \leq U_{max} \end{gathered} \tag{4.72}$$

The problem is to find the admissible control U that minimizes the time for which the system reaches the final state $\mathbf{X}_f$ from the initial one $\mathbf{X}_0$. Without loss of generality, and by normalization and shifting operations (if the two bounds of the control domain are not symmetric), the control is constrained to belong to a unit interval, i.e. $-1 \leq U \leq 1$.

To solve this problem, the Pontryagin Maximum Principle is applied to obtain necessary conditions for a reference trajectory of a system to be time-optimal: if $\mathbf{X}(t)$ is time-optimal trajectory defined on $[0, T]$, and $U(t)$ is the corresponding time-optimal reference control, then there exists an absolutely continuous vector function called the adjoint vector λ such that the following conditions are satisfied:

1. $\lambda(t) \neq 0 \; \forall t \in [0, T]$
2. $H(\mathbf{X}(t), \lambda(t), U^*(t)) = \min_U \mathbf{H}(\mathbf{X}(t), \lambda(t), U(t)) \leq 0$
3. The adjoint vector $\lambda(t)$ satisfies the following equation: $\dot{\lambda} = -\frac{\partial H}{\partial \mathbf{X}}$

A triple $(\mathbf{X}, \lambda, U)$ verifying the necessary conditions is called an extremal. First, consider the Hamiltonian H:

$$\begin{aligned} H(\mathbf{X}(t), \lambda(t), U(t)) &= 1 + \lambda^T f(X) \\ &= 1 + \lambda_1 (a_{11}\beta + a_{12}r + b_1\delta_r) + \lambda_2 (a_{21}\beta + a_{22}r + b_2\delta_r) \\ &\quad + \lambda_3 r + \lambda_4 V_e \cos(\beta + \psi) + \lambda_5 V_e \sin(\beta + \psi) \end{aligned} \tag{4.73}$$

and the co-state dynamics are given by:

$$\begin{aligned} \dot{\lambda_1} &= -\lambda_1 a_{11} - \lambda_2 a_{21} + \lambda_4 V_e \sin(\beta + \psi) - \lambda_5 V_e \cos(\beta + \psi) \\ \dot{\lambda_2} &= -\lambda_1 a_{12} - \lambda_2 a_{22} - \lambda_3 \\ \dot{\lambda_3} &= \lambda_4 V_e \sin(\beta + \psi) - \lambda_5 V_e \cos(\beta + \psi) \\ \dot{\lambda_4} &= 0 \\ \dot{\lambda_5} &= 0 \end{aligned} \tag{4.74}$$

Therefore, as λ_4 and λ_5 are constant on $[0, T]$ there exists $\mu = \sqrt{\lambda_4^2 + \lambda_5^2} \geq 0$ and $\zeta = \operatorname{atan2}(\frac{\lambda_5}{\sqrt{\lambda_4 + \lambda_5^2}}, \frac{\lambda_4^2}{\sqrt{\lambda_4^2 + \lambda_5^2}}) \in [0, 2\pi]$ such that $\forall t \in [0, T]$ then:

$$\begin{aligned} \dot{\lambda_1} &= -\lambda_1 a_{11} - \lambda_2 a_{21} + \mu V_e \sin(\beta + \psi - \zeta) \\ \dot{\lambda_2} &= -\lambda_1 a_{12} - \lambda_2 a_{22} - \lambda_3 \\ \dot{\lambda_3} &= \mu V_e \sin(\beta + \psi - \zeta) \\ \lambda_4 &= \mu \cos(\zeta) \\ \lambda_5 &= \mu \sin(\zeta) \end{aligned} \tag{4.75}$$

and the Hamiltonian becomes:

$$\begin{aligned} H(\mathbf{X}, \lambda, U) &= 1 + \lambda_1 (a_{11}\beta + a_{12}r) + \lambda_2(a_{21}\beta + a_{22}r) \\ &\quad + \lambda_3 r + \mu V_e \cos(\beta + \psi - \zeta) + (\lambda_1 b_1 + \lambda_2 b_2) \delta_r \end{aligned} \tag{4.76}$$

The minimization of the Hamiltonian with respect to the control δ_r is obtained by minimizing $\lambda^T g(X)\delta_r$. The control belongs to a unit control domain, then this min imization can be achieved by taking for δ_r the opposite sign of $\lambda^T g(X)$, then:

$$\begin{aligned} \delta_r &= 1 \quad \texttt{if } \lambda^T g < 0 \\ \delta_r &= -1 \quad \texttt{if } \lambda^T g > 0 \end{aligned}$$

The function $\Phi(t) = \lambda^T g$ defined along an extremal X, λ, δ_r is called the switching function associated to that system. The zeroes of this function are important for the study of optimal synthesis. If there exists a nonempty interval such that $\Phi(t)$ is identically zero, the extremal is singular on that interval. Assume now the extremal to be bang, i.e. takes its values in $\{-1, +1\}$ for almost a time t_s such that δ_r is not almost everywhere constant on any interval of the form $]t_s - \varepsilon, t_s + \varepsilon[$. t_s is called a switching time for δ_r and the corresponding state is called switching state. The trajectories corresponding to the controls $\delta_r = \pm 1$ are circles. The fastest way to turn is turning with the smallest radius, means that making the rudders deflection in one of its limits, i.e., to turn left or right [82, 84].

4.3.2.1 Singular Extremals

The singular controls are characterized by the fact that $\Phi(t)$ is identically zero in a nonempty interval. However, the Pontryagin Maximum Principle loses its discriminative nature, i.e. every controls in **U** satisfy the necessary conditions. In this case, some additional conditions are needed. The fact that $\Phi(t) = 0$ in a nonempty interval implies that all its time derivatives are null in that interval, i.e.

$$\dot{\Phi}(t) = \ddot{\Phi}(t) = \ldots = \Phi^{2m}(t) = 0 \tag{4.77}$$

the process of derivation is stopped when the control appear in the expression of these derivation. For an affine system,

$$\Phi^{2m}(t) = a(\mathbf{X}, \lambda) + b(\mathbf{X}, \lambda)\delta_r = 0 \tag{4.78}$$

k is called the order of singular control. Hence, the singular control can be expressed as:

$$\delta_r = -\frac{a(\mathbf{X}, \lambda)}{b(\mathbf{X}, \lambda)} \tag{4.79}$$

Proposition 4.1 *The singular controls of this system are of the first order and are never abnormal.*

Proof Let's derive the switching function $\phi(t)$:

$$\Phi(t) = \lambda^T g = b_1\lambda_1 + b_2\lambda_2 \tag{4.80}$$

Differentiation gives

$$\begin{aligned} \dot{\Phi}(t) &= b_1\dot{\lambda}_1 + b_2\dot{\lambda}_2 \\ \ddot{\Phi}(t) &= b_1\ddot{\lambda}_1 + b_2\ddot{\lambda}_2 \end{aligned} \tag{4.81}$$

Nullifying these equations, the following relation is obtained:

$$\lambda_2 = -\frac{b_1}{b_2}\lambda_1 \tag{4.82}$$

while replacing in the first and the second equations of the system of equations (4.74), we find:

$$\begin{aligned}\dot{\lambda}_1 &= -\left(a_{11} - \frac{b_1}{b_2}a_{21}\right)\lambda_1 + \dot{\lambda}_3 \\ \dot{\lambda}_2 &= -\left(a_{22} - \frac{b_2}{b_1}a_{12}\right)\lambda_2 - \lambda_3\end{aligned} \tag{4.83}$$

□

Proposition 4.2 *The system of differential Eqs.* (4.82, 4.83) *has as solution* $\lambda_1 = \lambda_2 = \lambda_3 = 0$.

Proof The condition $\lambda_1 = \lambda_2 = \lambda_3 = 0$ gives $\sin(\psi + \beta - \zeta) = 0$ implying $\psi + \beta - \zeta = k\pi$ with $k \in \mathbf{Z}$. This result represents the necessary condition for the existence of the singular control. So the Hamiltonian becomes:

$$H(X, \lambda, u) = 1 + \mu V_e \cos(\beta + \psi - \zeta) = 0 \tag{4.84}$$

from this,

$$\mu = -\frac{1}{V_e \cos(\beta + \psi - \zeta)} = \pm\frac{1}{V_r} \tag{4.85}$$

implying that μ is never equal to zero, because the zero value of μ implies a null adjoint vector λ, contradicting Pontryagin Maximum Principle statements.

To prove the minimality of the singular controls, the generalized convexity condition often called, strengthened Legendre-Clebsh condition must be tested:

$$(-1)^k \frac{\partial}{\partial u}\left(\frac{d^{2k}\Phi}{dt^{2k}}\right) \geq 0 \tag{4.86}$$

k is the order of singular control, thus:

$$(-1)^k \frac{\partial}{\partial \delta_r}\left(\frac{d^{2k}\Phi}{dt^{2k}}\right) = -b_1^2 \mu V_e \cos(\psi + \beta - \zeta) = b_1^2 \geq 0 \tag{4.87}$$

Hence, the singular control can be given by:

$$\delta_r = \left(a_{11} + \frac{b_2}{b_1}(1 + a_{12})\right)\tan(\psi + \beta - \zeta) - \frac{a_{11}\beta + (1 + a_{12}r)}{b_1} \tag{4.88}$$

from the existence condition of the singular control.

Once the singular control is determined, the geometric shape of the reference trajectory of the airship under this singular control is illustrated. From the singular control necessary condition, $\psi + \beta - \zeta$ is constant, and ζ is constant too, thus $\psi + \zeta$ is constant. This implies that the angle between the relative velocity V_e and the x reference axis is constant. Thus the singular trajectory is a straight line. The same result can be found by applying the control defined below in the dynamics of:

$$\dot{\beta} + \dot{\psi} = 0 \tag{4.89}$$

The reference trajectory is thus a line. Transition from a curved path to the line requires the application of the singular control which grows monotonously. This explosion of control is due to the rejection of the discontinuities in path curvature by the lighter than air dynamics. For this reason this curvature has to be smoothed by optimizing the transition time between a non zero curvature trajectory and a straight line. When two configurations are aligned the optimal path is obviously a straight line, i.e. $\beta = r = 0$, which correspond to $\delta_r^* = 0$. □

4.3.2.2 Optimal Transition of a Non Null Curvature Path to a Straight Line

The switch between the non zero curvature path and a straight line must be characterized separately. The shapes of the optimal state and control trajectories in each mode separately have been already characterized. Therefore, the time interval corresponding to the mode switching need to be specified and the parameters that determine when and for how long the singular control lasts. Both issues are addressed using a continuity argument. The line is characterized by zero values of β, r, and the non null curvature paths are characterized by non zero values of β, r. The dynamics of β, r are used for achieving this objective. In optimal control literature, the following theorem is demonstrated.

Proposition 4.3 *For any linear normal system the optimal control is of bang-bang type.*

Proof The normality condition means that the system is controllable with respect to each of its control inputs. The determinant of the controllability matrix is:

$$a_{21}b_1^2 + b_1 b_2 (a_{22} - a_{11}) - a_{12}b_2^2$$

Let's find the switching surface allowing the system to intersect the origin, starting from any initial condition within this surface, and under a specific control, i.e. $\delta = \pm 1$. The dynamics of the yaw rate and the side-slip angle are asymptotically stable. To simplify the computations, a separation of dynamics is performed by state matrix diagonalization, using linear state space variable transformation. Let

$$\begin{pmatrix} z_1 \\ z_2 \end{pmatrix} = T^{-1} \begin{pmatrix} \beta \\ r \end{pmatrix} \tag{4.90}$$

where T is a 2×2 matrix formed by the eigenvectors corresponding to the system eigenvalues. The resultant diagonalized dynamics are given by:

$$\begin{aligned} \dot{z}_1 &= \lambda_1 z_1 + b_{z1}\delta_r \\ \dot{z}_2 &= \lambda_2 z_2 + b_{z2}\delta_r \end{aligned} \tag{4.91}$$

The corresponding solutions of this system are given by:

$$
\begin{aligned}
z_1 &= \frac{b_{z1}\delta_r + z_{10}\lambda_1}{\lambda_1}\exp^{\lambda_1 t} - b_{z1}\frac{\delta_r}{\lambda_1} \\
z_2 &= \frac{b_{z2}\delta_r + z_{20}\lambda_2}{\lambda_2}\exp^{\lambda_2 t} - b_{z2}\frac{\delta_r}{\lambda_2}
\end{aligned} \tag{4.92}
$$

where z_{10} and z_{20} are the initial conditions. The cross time with the origin t_c is easily calculated as:

$$
t_c = \frac{\ln\left(b_{z1}\delta_r/(b_{z1}\delta_r + z_{10}\lambda_1)\right)}{\lambda_1} = \frac{\ln\left(b_{z2}\delta_r/(b_{z2}\delta_r + z_{20}\lambda_1)\right)}{\lambda_2} \tag{4.93}
$$

Thus, the following relation must hold

$$
z_{10} = -\delta_r \frac{(b_{z2}\delta_r/(b_{z2}\delta_r + z_{20}\lambda_2))^{\lambda_2/\lambda_1} - 1}{(b_{z2}\delta_r/(b_{z2}\delta_r + z_{20}\lambda_2))^{\lambda_2/\lambda_1}} \tag{4.94}
$$

This equation defines the switching surface in z_1 and z_2 coordinates corresponding to $\delta_r = \pm 1$.

The algorithm of the optimal transition is based on the detection of the cross points of the bang-bang trajectories with the switching surface, the idea is to compare the horizontal distance of the system state point (z_1, z_2) from the switching surface by replacing the z_1 and z_2 in Eq. (4.94) and switching when the condition $|z_1 - z_{10}| \leq \varepsilon$ is hold. In this case the control switch to the other bang. □

4.4 Zermelo Navigation Problem

Zermelo's problem was originally formulated to find the quickest nautical path for a ship at sea in the presence of currents, from a given departure point in $\Re^2$ to a given destination point. It can also be applied to the particular case of a lighter than air robot with a zero flight path angle and the wind velocity represented by $W = (W_x, W_y)$ [92, 96, 147, 166].

4.4.1 Navigation Equation

Time optimal trajectory generation can be formulated as follows:

$$
\min \int_0^T dt \tag{4.95}
$$

subject to

$$
\begin{aligned}
\dot{x} &= u_1(t) + W_x \\
\dot{y} &= u_2(t) + W_y
\end{aligned} \tag{4.96}
$$

with the constraint $u_1^2(t) + u_2^2(t) \leq V_{max}^2$ If the terminal point is reachable at any time, then it is reachable in the minimal time. However, if the wind is too strong, there may be points that are not reachable at all.

The Hamiltonian is classically given by: $H = 1 + \lambda_1(u_1(t) + W_x) + \lambda_2(u_2(t) + W_y)$ where the Lagrange multipliers are represented by λ_1, λ_2. The application of the necessary condition of optimality

$$\begin{aligned} \frac{d\lambda_1}{dt} &= -\frac{\partial H}{\partial x} = -\lambda_1 \frac{\partial W_x}{\partial x} - \lambda_2 \frac{\partial W_y}{\partial x} \\ \frac{d\lambda_2}{dt} &= -\frac{\partial H}{\partial y} = -\lambda_1 \frac{\partial W_x}{\partial y} - \lambda_2 \frac{\partial W_y}{\partial y} \end{aligned} \tag{4.97}$$

Each extremal control $u^*(t)$ must satisfy $\|u^*(t)\| = V_{max}$ for almost all t. The maximality condition yields that

$$u^*(t) = V_{max} \frac{\lambda(t)}{\|\lambda\|}$$

for almost all t, as $\lambda(t)$ cannot be identically zero.

Zermelo's navigation formula consists of a differential equation for $u^*(t)$ expressed in terms of only the drift vector and its derivatives. The derivation can be explained as follows. Let the angle $\mu(t)$ given by $u_1(t) = V_{max} \cos \mu(t)$ and $u_2(t) = V_{max} \sin \mu(t)$ then

$$\cos \mu(t) = \frac{\lambda_1}{\|\lambda\|}, \qquad \sin \mu(t) = \frac{\lambda_2}{\|\lambda\|}$$

Differentiating these relations, the following equalities can be given:

$$\begin{aligned} \cos \mu \frac{d\|\lambda\|}{dt} - \|\lambda\| \sin \mu \frac{d\mu}{dt} &= \frac{d\lambda_1}{dt} = -\lambda_1 \frac{\partial W_x}{\partial x} - \lambda_2 \frac{\partial W_y}{\partial x} \\ \sin \mu \frac{d\|\lambda\|}{dt} + \|\lambda\| \cos \mu \frac{d\mu}{dt} &= \frac{d\lambda_2}{dt} = -\lambda_1 \frac{\partial W_x}{\partial y} - \lambda_2 \frac{\partial W_y}{\partial y} \end{aligned} \tag{4.98}$$

Finally the Zermelo navigation equation is given by:

$$\frac{d\mu}{dt} = -\cos^2 \mu \frac{\partial W_x}{\partial y} + \sin \mu \cos \mu \left(\frac{\partial W_x}{\partial x} - \frac{\partial W_y}{\partial y} \right) + \sin^2 \mu \frac{\partial W_x}{\partial x} \tag{4.99}$$

4.4.2 One Particular Solution

When the problem is to find minimum-time paths through a 2D region of position-dependent vector velocity [92]:

$$\begin{aligned} \dot{x} &= V \sin \chi + W_x(x, y) \\ \dot{y} &= V \cos \chi + W_x(x, y) \end{aligned} \tag{4.100}$$

The heading angle is the control available for achieving the minimum time objective.

If $W_x(x,y) = -V_V y$ and $W_y(x,y) = 0$, it has been proved in [92] that

$$y = \frac{V}{V_w}\left(\frac{1}{\sin\chi} - \frac{1}{\sin\chi_f}\right)$$
$$x = \frac{V}{2V_W}\left(\operatorname{asinh}(\tan\chi_f) - \operatorname{asinh}(\tan\chi) + \tan\chi\left(\frac{1}{\sin\chi} - \frac{1}{\sin\chi_f}\right) - \frac{1}{\cos\chi_f}\left(\tan\chi_f - \tan\chi\right)\right) \tag{4.101}$$

The time to go is given by

$$T = \frac{1}{V_W}\left(\tan\chi_f - \tan\chi\right) \tag{4.102}$$

4.5 3D Trajectory Design with Wind

The contribution of this section is the analysis of time-optimal 3D trajectories for a lighter than air robot in a constant velocity wind. This section studies both issues that arise in the planning of trajectory of lighter than air robot in winds: the controllability of the lighter than air robot expressed by its kinematical model and time optimal trajectory characterization in 3D space.

The problem of transferring a dynamical system from an arbitrary initial configuration to a desired target in minimum time is of fundamental interest as an optimal control problem. Trajectory plan for a lighter than air robot must incorporate wind as a significant factor that can affect both the feasibility and optimality of trajectory. A family of primitives is deduced from the resolution of this optimization problem.

In guidance studies, one assumes that only local information on the wind flowfield is available and determines a near optimal trajectory, namely a trajectory that approximates the behavior of the optimal trajectory [148].

4.5.1 Determination of the Reference Controls

The kinematic model of the lighter than air robot exhibits a property known as differential flatness particularly relevant in planning problems. A nonlinear system

$$\dot{\mathbf{X}} = f(\mathbf{X}) + g(\mathbf{X})\mathbf{U} \tag{4.103}$$

is differentially flat if there exists a set of outputs $\mathbf{Y}$ called flat outputs such that the state $\mathbf{X}$ and the control inputs $\mathbf{U}$ can be expressed algebraically as a function of $\mathbf{Y}$ and its time derivative up to a certain order:

$$\mathbf{X} = \mathbf{X}\left(\mathbf{Y}, \dot{\mathbf{Y}}, \ddot{\mathbf{Y}}, \ldots, \mathbf{Y}^{(r)}\right)$$
$$\mathbf{U} = \mathbf{U}\left(\mathbf{Y}, \dot{\mathbf{Y}}, \ddot{\mathbf{Y}}, \ldots, \mathbf{Y}^{(r)}\right) \tag{4.104}$$

As a consequence, once an output is assigned for **Y**, the associated trajectory of the state **X** and history of control inputs are uniquely determined. The dynamic model is recalled as:

$$\begin{aligned}\dot{x} &= V \sin\chi \cos\gamma + W_x \\ \dot{y} &= V \cos\chi \cos\gamma + W_y \\ \dot{z} &= V \sin\gamma + W_z\end{aligned} \tag{4.105}$$

$$\begin{aligned}\dot{V} = &\frac{1}{m + m_{11}}(T\cos\alpha - D + (B - mg)\sin\gamma) \\ &- \frac{m}{m + m_{11}}\left(\dot{W}_x \cos\gamma \sin\chi + \dot{W}_y \cos\gamma \cos\chi + \dot{W}_z \sin\gamma\right)\end{aligned} \tag{4.106}$$

$$\begin{aligned}\dot{\chi} = &\frac{1}{(m + m_{22})V\cos\gamma}(L + T\sin\alpha)\sin\sigma \\ &- \frac{m}{m + m_{22}}\left(\frac{\dot{W}_x \cos\chi - \dot{W}_y \sin\chi}{V\cos\gamma}\right)\end{aligned} \tag{4.107}$$

$$\begin{aligned}\dot{\gamma} = &\frac{-1}{(m + m_{33})V}(L\cos\sigma + T\cos\sigma\sin\alpha + (B - mg)\cos\gamma) \\ &+ \frac{m}{(m + m_{33})V}\left(\dot{W}_x \sin\gamma \sin\chi + \dot{W}_y \sin\gamma \cos\chi + \dot{W}_z \cos\gamma\right)\end{aligned} \tag{4.108}$$

If one assumes that the control inputs are T, σ, α or equivalently T, σ, L then the lighter than air motion is differentially flat. By determining a suitable lighter than air robot trajectory in Cartesian coordinates x, y, and z, the required lighter than air robot controls can be calculated.

$$V = \sqrt{(\dot{x} - W_x)^2 + (\dot{y} - W_y)^2 + (\dot{z} - W_z)^2} \tag{4.109}$$

$$\gamma = \operatorname{asin}\frac{\dot{z} - W_z}{V} \tag{4.110}$$

$$\chi = \operatorname{atan}\frac{\dot{x} - W_x}{\dot{y} - W_y} \tag{4.111}$$

Using the following notations,

$$\begin{aligned}\eta_1 = &(m + m_{11})\dot{V} - (B - mg)\sin\gamma \\ &+ m\left(\dot{W}_x \cos\gamma \sin\chi + \dot{W}_y \cos\gamma \cos\chi + \dot{W}_z \sin\gamma\right) \\ \eta_2 = &(m + m_{22})V\dot{\chi}\cos\gamma + m\left(\dot{W}_x \cos\chi - \dot{W}_y \sin\chi\right)\end{aligned} \tag{4.112}$$

$$\eta_3 = -(m+m_{33})V\dot{\gamma} - (B-mg)\cos\gamma$$
$$- m\left(\dot{W}_x \sin\gamma \sin\chi + \dot{W}_y \sin\gamma \cos\chi + \dot{W}_z \cos\gamma\right)$$

The bank angle is given by:

$$\sigma = \operatorname{atan}\frac{\eta_2}{\eta_3} \tag{4.113}$$

While the thrust is given by:

$$T = \sqrt{\left(\eta_1 + L\frac{C_D}{C_L}\right)^2 + (\eta_2 - L\sin\sigma)^2 + (\eta_3 - L\cos\sigma)^2} \tag{4.114}$$

The lift is given by resolving the following equation:

$$L^2 - (\eta_2 - T\sin\alpha\sin\sigma)^2 + (\eta_3 - T\sin\alpha\cos\sigma)^2 = 0 \tag{4.115}$$

where the lift force L is normal to the velocity vector of the lighter than air robot with respect to the air and is contained in the plane of symmetry of the vehicle. The drag force D is parallel and in the opposite direction of the velocity vector.

$$D = L\frac{C_D}{C_L} = L\left(\frac{C_{D0}}{C_l} + KC_L\right) \tag{4.116}$$

The constraints are first expressed in terms of thrust and velocities and then transformed into limitations on flight path and heading angles.

Remark 4.2 As the lighter than air robot admits a set of flat outputs, these may be exploited to solve planning problems efficiently. In fact, some interpolation schemes can be used to solve the path of **Y** in such a way to satisfy the appropriate boundary conditions. The evolution of the other configuration variables, together with the associated control inputs, can then be computed algebraically from **Y**. The resulting configuration space path will automatically satisfy the nonholonomic constraints.

4.5.2 Accessibility and Controllability

The nonlinear dynamics model can be written as a set of first-order ordinary differential equation as:

$$\begin{aligned} \dot{X} &= \mathbf{f}(X,U) = f(X) + g(X)U \\ Y &= h(X) \end{aligned} \tag{4.117}$$

where X is the state vector $X \in R^n$, U is the control vector $U \in R^m$, Y is the measured output $Y \in R^p$.

Remark 4.3 As modeling involves approximation and some degree of uncertainty in dealing with dynamic systems, properties whose validity in nominal situations may imply validity in almost all situations are called generic properties. These are properties that hold on open and dense subsets of suitable domains of definition, provided they hold at some points of such domains.

A basic notion is that of reachable state and controllability. Controllability concerns the possibility of steering the system from a state X_0 to another state X_1. For linear systems, controllability is a structural property, in the sense that any linear system can be split into a controllable subsystem and an autonomous uncontrollable one [204].

Definition 4.1 A system of the form (4.117) is said to be controllable at X_0 if there exists a neighborhood V of X_0 such that any state $X_1 \in V$ is reachable from X_0.

The notion of the observability of a system concerns the possibility of recovering the state $X(t)$ from knowledge of the measured output $Y(t)$, the input $U(t)$ and possibly a finite number of their time derivatives $Y^{(k)}(t)$, $k \geq 0$ and $U^{(l)}$, $l \geq 0$. The structural property which can be easily characterized in a nonlinear framework concerns the existence of an open and dense submanifold of the state space R^n around whose points the system is locally observable. The use of an observer that evaluates the state from the knowledge of inputs and outputs is in order whenever the state itself is not directly measurable, but its value is required for computing a feedback or for monitoring the system behavior. In contrast to the linear situation, observability of a given nonlinear system is necessary but not sufficient to assure the possibility of constructing an observer.

Controllability is an important notion for affine systems with or without drift. Sussmann [203] and Jurdjevic [96] introduced the theory of Lie groups and their associated Lie algebras into the context of nonlinear control to express notions such as controllability, observability and realization theory. Some of the early works on nonlinear controllability of driftless systems were based on linearization of nonlinear systems. It was observed that if the linearization of a nonlinear system at an equilibrium point is controllable, the system itself is locally controllable. Later, a differential geometric approach to the problem was adopted in which a control system was viewed as a family of vector fields. It was observed that a lot of the interesting control theoretic information was contained in the Lie brackets of these vector fields.

Driftless nonholonomic control systems have been extensively studied in recent years. Chow's theorem [96] leads to the characterization of controllability for systems without drift. It provides a Lie algebra rank test, for controllability of nonlinear systems without drift, similar in spirit to that of Kalman's rank condition for linear systems. In the setting of controlled mechanical systems, the Lagrangian dynamics, being second order, necessarily include drift. In this setting, Chow's theorem cannot be used to conclude controllability. Studying controllability of general systems with drift is usually a hard problem. The discussion of nonholonomic system with drift in the literature has been concentrated on the so-called dynamic extension of drift-free systems with the addition of integrators. Sufficient conditions for the controllability of a conservative dynamical nonlinear affine control system on a compact Riemannian manifold are presented, if the drift vector field is assumed to be weakly positively Poisson stable [47, 49, 181, 218].

Let's begin with a brief study of controllability of nonlinear systems applied to this system:

$$\begin{aligned} \dot{x} &= V \sin\chi \cos\gamma + W_x \\ \dot{y} &= V \cos\chi \cos\gamma + W_y \\ \dot{z} &= V \sin\gamma + W_z \\ \dot{V} &= u_1 \\ \dot{\chi} &= u_2 \\ \dot{\gamma} &= u_3 \end{aligned} \tag{4.118}$$

or in an affine nonlinear control system with drift:

$$\dot{X} = f(X) + g_1 u_1 + g_2 u_2 + g_3 u_3 = A(X)X + W + G(X)U \tag{4.119}$$

where the state variable is defined as $X = (x, y, z, V, \chi, \gamma)^T$, the control variable by $U = (\dot{V}, \dot{\chi}, \dot{\gamma})^T$, the drift function by:

$$\begin{aligned} f(X) &= \begin{pmatrix} V \sin\chi \cos\gamma + W_x \\ V \cos\chi \cos\gamma + W_y \\ V \sin\gamma + W_z \\ 0 \\ 0 \\ 0 \end{pmatrix} \\ &= \begin{pmatrix} 0 & 0 & 0 & \cos\gamma \sin\chi & 0 & 0 \\ 0 & 0 & 0 & \cos\gamma \cos\chi & 0 & 0 \\ 0 & 0 & 0 & \sin\gamma & 0 & 0 \\ 0 & 0 & 0 & 0 & 0 & 0 \\ 0 & 0 & 0 & 0 & 0 & 0 \\ 0 & 0 & 0 & 0 & 0 & 0 \end{pmatrix} \begin{pmatrix} x \\ y \\ z \\ V \\ \chi \\ \gamma \end{pmatrix} + \begin{pmatrix} W_x \\ W_y \\ W_z \\ 0 \\ 0 \\ 0 \end{pmatrix} \end{aligned} \tag{4.120}$$

and the input functions are given by:

$$\begin{aligned} g_1 &= (0, 0, 0, 1, 0, 0)^T \\ g_2 &= (0, 0, 0, 0, 1, 0)^T \quad \text{or} \quad G(X) = [g_1\ g_2\ g_3] \\ g_3 &= (0, 0, 0, 0, 0, 1)^T \end{aligned} \tag{4.121}$$

Remark 4.4 This control system with drift

$$\dot{X} = f(X) + \sum_{i=1}^{m} g_i(X) U_i$$

may be thought as a special case of the drift free system by defining the new system

$$\dot{X} = \sum_{i=0}^{m} g_i(X)U_i$$

with the input U_0 attached to the vector field $g_0(X) = f(X)$ frozen at 1.

Remark 4.5 The Lie bracket is defined as $[f, g] = \frac{\partial g}{\partial X} f - \frac{\partial f}{\partial X} g$.

Definition 4.2 The controllability Lie algebra C of an affine control system is defined to be the span over the ring of smooth functions of elements of the form $[X_k, [X_{k-1}, [\dots, [X_2, X_1], \dots]]]$ where $X_i \in \{f, g_1, \dots, g_m\}$ and $k = 0, 1, 2, \dots$

A Lie subalgebra of a module is a submodule that is closed under the Lie bracket. The controllability Lie algebra may be used to define the accessibility distribution as the distribution

$$C(X) = \text{span}\,\{\mathbf{X}(X), \mathbf{X} \in \mathbf{C}\}$$

From the definition of C, it follows that $C(X)$ is an involutive distribution. The controllability Lie algebra is the smallest Lie algebra containing $\{f, g_1, \dots, g_m\}$ which is closed under Lie bracketing. The controllability distribution contains valuable information about the set of states that are accessible from a given starting point X_0.

Definition 4.3 Let $R^{\nu}(X_0, T) \subset R^m$ be the subset of all states accessible from state X_0 in time T with the trajectories being confined to a neighborhood U of X_0. This is called the reachable set from X_0.

Several important results have been derived based on the structure of the Lie algebra generated by the control vector fields. Assume $X \in M \subset \Re^6$ where M is a smooth manifold. Let $X(t, X_0, u)$ denote the solution for $t \geq 0$ for a particular input function u and initial condition $X(0) = X_0$.

Let $R^{\nu}(X_0, T) = \{X \in M$, there exists an admissible input $u : [0, T] \to u$ such that

$$X(t, X_0, u) \in V, \quad 0 \leq t \leq T \quad \text{and} \quad X(T) = X_f\} \tag{4.122}$$

By definition, all the Lie brackets that can be generated using these vector fields belong to A. The accessibility algebra A of the system (4.118) is the smallest sub algebra of $V \subset \Re^n$ that contains f, g_1, g_2, g_3. The accessibility distribution Δ_A of the system (4.118) is defined as: $\Delta_A = \text{span}\{\nu | \nu \in A\}$, thus Δ_A is the involutive closure of $\Delta = \text{span}\{f, g_1, g_2, g_3\}$. The computation of Δ_A may be organized as an iterative procedure:

$$\Delta_A = \text{span}\left\{\nu \middle| \nu \in \Delta_i; \forall i \geq 1\right\} \tag{4.123}$$

With:

$$\Delta_1 = \Delta = \text{span}\{f, g_1, g_2, g_3\}$$
$$\Delta_i = \Delta_{i-1} + \text{span}\left\{[g, v], g \in \Delta_1, v \in \Delta_{i-1}\right\}, \quad i \geq 2$$

This procedure stops after K steps, where K is the smallest integer such that $\Delta_{K+1} = \Delta_K > \Delta_A$. This number is called the nonholonomy degree of the system and is related to the 'level' of Lie brackets that must be included in Δ_A.

Since $\dim(\Delta_A) \leq n$, so the following condition is necessary:

$$K \leq n - m$$

Let us now find Δ_A of this system (4.118):

$$\Delta_1 = \text{span}\{f, g_1, g_2, g_3\}$$

At the second level, the following relationship can be written:

$$\Delta_2 = \Delta_1 + \text{span}\left\{ \begin{array}{l} g_4 = [f, g_1], g_5 = [f, g_2], g_6 = [f, g_3], \\ g_7 = [g_1, g_2], g_8 = [g_1, g_3], g_9 = [g_2, g_3] \end{array} \right\} \tag{4.124}$$

with

$$[f, [f, g_1]] = [f, [f, g_2]] = [f, [f, g_3]] = [g_3, [f, g_3]] = 0_{6\times 1}$$
$$[g_1, [f, g_1]] = V [f, g_3]$$
$$[g_1, [f, g_3]] = -[g_3, [f, g_1]] = -V [f, g_1]$$
$$[g_1, [f, g_2]] = -[g_2, [f, g_1]] = -\tan\gamma [f, g_2]$$
$$[g_2, [f, g_3]] = -[g_3, [f, g_2]] = -V [f, g_2]$$
$$[g_2, [f, g_2]] = (V\cos\chi\cos\gamma \;\; V\sin\chi\cos\gamma \;\; 0\;0\;0\;0)^T$$

Identical calculations are made for level 3 and 4 to obtain

$$\Delta_4 = \Delta_3 = \Delta_A = \text{span}\left\{f, g_1, g_2, g_3, [f, g_1], [f, g_2], [f, g_3]\right\} \tag{4.125}$$

The Lie brackets are zero since level 2, thus this system is nilpotent.

The rank of the matrix obtained: $M = (f, g_1, g_2, g_3, [f, g_1], [f, g_2])$

Straightforward calculations allow to write:

$$\text{rank}(M)$$
$$= \text{rank}\begin{pmatrix} V\sin\chi\cos\gamma + W_x & 0 & 0 & 0 & V\sin\chi\sin\gamma & V\cos\chi\cos\gamma \\ V\cos\chi\cos\gamma + W_y & 0 & 0 & 0 & V\cos\chi\sin\gamma & -V\sin\chi\cos\gamma \\ V\sin\gamma + W_z & 0 & 0 & 0 & -V\cos\gamma & 0 \\ 0 & 1 & 0 & 0 & 0 & 0 \\ 0 & 0 & 1 & 0 & 0 & 0 \\ 0 & 0 & 0 & 1 & 0 & 0 \end{pmatrix} \tag{4.126}$$

where the determinant of this matrix is given by:

$$\det(M) = -V^3 \cos\gamma - W_z V^2 \sin\gamma \cos\gamma - W_x V^2 \sin\chi \cos^2\gamma \\ - W_y V^2 \cos\chi \cos^2\gamma \tag{4.127}$$

So the previous relation (4.127) must be studied, verifying that the determinant is different from zero. With this condition, system (4.118) verifies the Lie Algebra rank condition and is locally accessible.

Therefore the non-holonomy degree of the system is: $(K = 3)$ for $\det(M) \neq 0$ and its satisfies the condition $(K = 3) \leq (n - m = 3)$. If the drift f is positively Poisson stable vector field, and as the accessibility Lie algebra rank condition is satisfied, the system is controllable.

The pointwise linear controllability means that the pair $[A(X)G(X)]$ is controllable for any $X \in \Re^6$, with a condition

$$\det\left[G(X)\ A(X)G(X)\ A(X)^2G(X)\ A(X)^3G(X)\ A(X)^4G(X)\ A(X)^5G(X)\right] > 0$$

with the following property

$$A(X)^2G(X) = A(X)^3G(X) = A(X)^4G(X) = A(X)^5G(X) = \mathbf{0}_{6\times 6}$$

Finally, the following result is obtained:

$$\mathrm{rank}\left[G(X)\ A(X)G(X)\ A(X)^2G(X)\ A(X)^3G(X)\ A(X)^4G(X)\ A(X)^5G(X)\right] = 4$$

for $\gamma \neq 0$ and $\chi \neq 0$.

4.5.3 Motion Planning when Wind Can Be Neglected

Nonholonomic motions planning relies on finding a trajectory in the state space between given initial and final configurations subject to nonholonomic constraints. The Lie algebraic method relies on a series of local planning around consecutive current states. Global trajectory results from joining local trajectories. At a current state, a direction of motion toward the goal state is established. Then, a rich enough space of controls is taken. As the system is controllable, via some vector fields, the controls are able to generate the fields [67, 181].

The steering method for affine driftless systems exploits different properties of such a system namely nilpotence, chained form and differential flatness.

1. Nilpotence: A control system is nilpotent as soon as the Lie brackets of the control vector fields vanish from some given length. For such system, it is possible to compute a basis of the control Lie algebra from a Philip Hall family. Because the system is nilpotent, each exponential of Lie brackets can be developed exactly as finite combination of the control vector fields and thus find piecewise constant controls steering the system exactly to the goal

2. Differential flatness is a property enjoyed by this mechanical system. A flat system is a system such that there exists a finite set of variable y_i differentially independent which appear as differential function of the system variables (state variables and inputs) and a finite number of their derivatives, each system variable being itself a function of the y_i and a finite number of their derivatives. The variables y_i are called the linearizing outputs of the system. In this case, path planning takes place in a space defined by parameters different in general from the configuration variables
3. Chained forms: For systems that can be converted into chained form, it is possible to steer them exactly to the goals using either sinusoidal, polynomial or piecewise constant controls.

As compared with driftless systems, relatively few approaches have so far been proposed for the stabilization of systems with drift. The difficulty in steering systems with drift arises from the fact, that in the most general case of non recurrent or instable drift, the system motion along the drift vector needs to counteracted by enforcing system motions along adequately chosen Lie bracket vector fields in the system, underlying controllability Lie algebra.

This system Ξ

$$\dot{X} = f(X) + g_1u_1 + g_2u_2 + g_3u_3 = A(X)X + W + G(X)U \tag{4.128}$$

can be steered along these Lie brackets $\{g_1, g_2, g_3, g_4, g_5, g_6\}$. The vectors g_4, g_5, g_6 represent new motion directions that can be followed approximately. Locally generating motion in these directions is slower than following the vector fields g_1, g_2, g_3. The Philip Hall basis gives a way to choose the smallest number of Lie products that must be considered to generate a basis for this distribution.

The level of difficulty in steering controllable systems is proportional to the order of Lie brackets. When the dimension of the L is 6 according to Chow's theorem and Lie algebra rank condition, the system is proved to be completely nonholonomic and controllable.

The motion planning problem is concerned with finding a control that steers the system from an initial configuration X_i to the final configuration $X_f \in \Re^5$ along a certain trajectory. The following extension Ξ_e of Eq. (4.128) is proposed by adding higher order Lie bracket motions

$$\dot{X} = f(X) + g_1u_1 + g_2u_2 + g_3u_3 + g_4u_4 + g_5u_5 + g_6u_6 \tag{4.129}$$

where u_4, u_5, u_6 are fictitious inputs that may not correspond to the actual system inputs. It has been proved above that the system is nilpotent, as all the Lie brackets vanish from level 3. Once the extended system has been selected, the solution proposed here of the motion planning problem involves two steps:

1. STEP 1: Find a control $U = (u_1, u_2, u_3, u_4, u_5, u_6)^T$ that steers Ξ_e from a point X_i to a point X_f, defining a curve $\varpi(t) : [0, T] \rightarrow \Re^5$, then the tangent vector $\dot{\varpi}(t)$ is expressed as a linear combination of

$$g_1(\varpi(t)), \qquad g_2(\varpi(t)), \qquad g_3(\varpi(t)),$$

$$g_4(\varpi(t)), \qquad g_5(\varpi(t)), \qquad g_6(\varpi(t))$$

the corresponding coefficients being $u_i(t), i = 1..6$

2. STEP 2: Use U to compute a control u that steers Ξ_e from the point X_i to a point X_f by computing the Philip Hall coordinates of the extended system and by finding a control u from the Philip Hall coordinates.

As $\{g_1, g_2, g_3, g_4, g_5, g_6\}$ are vector fields defined in a neighborhood N of a point X such that at each point of N, the set $\{g_1, g_2, g_3, g_4, g_5, g_6\}$ constitutes a basis of the tangent space, then there is a smaller neighborhood of X on which the maps

$$(\alpha_1, \alpha_2, \alpha_3, \alpha_4, \alpha_5) \rightarrow e^{\alpha_1 g_1 + \alpha_2 g_2 + \alpha_3 g_3 + \alpha_4 g_4 + \alpha_5 g_5 + \alpha_6 g_6} X$$

and

$$(\alpha_1, \alpha_2, \alpha_3, \alpha_4, \alpha_5) \rightarrow e^{\alpha_1 g_1} e^{\alpha_2 g_2} e^{\alpha_3 g_3} e^{\alpha_4 g_4} e^{\alpha_5 g_5} e^{\alpha_6 g_6} X$$

are two coordinate systems, called the first and second normal coordinate system associated to $\{g_1, g_2, g_3, g_4, g_5, g_6\}$. The Campbell Baker Hausdorff Dynkin (CBHD) formula states precisely the difference between the two systems for a sufficiently small τ, one has

$$e^{\tau f}.e^{\tau g} = e^{\tau f + \tau g - 0.5[f,g] + \tau^2 \varepsilon(t)}$$

where $\varepsilon(\tau) \rightarrow 0$ when $\tau \rightarrow 0$.

Actually, the whole formula gives an explicit form for the $\varepsilon(t)$ function. More precisely, $\varepsilon(t)$ yields a formal series whose coefficients c_k of τ^k are combinations of brackets of degree k, i.e.

$$\tau^2 \varepsilon(t) = \sum_{k=3}^{\infty} \tau^k c_k$$

Roughly speaking, the Campbell Baker Hausdorff Dynkin formula tells how a small time controllable nonholonomic system can reach any point in the neighborhood of a starting point. The formula is the hard core of the local controllability concepts. It yields a method for explicitly computing admissible paths in a neighborhood of a point.

4.5.3.1 Philip Hall Coordinates

The Philip Hall basis of the controllability Lie Algebra generated by $\{f, g_1, g_2, g_3\}$ is defined as follows $\{g_1, g_2, g_3, g_4, g_5, g_6\}$. All flows can be represented in the form

$$\begin{aligned} S(t) &= e^{h_6(t)g_6} e^{h_5(t)g_5} e^{h_4(t)g_4} e^{h_3(t)g_3} e^{h_2(t)g_2} e^{h_1(t)g_1} \\ S(t) &= e^{\overline{h}_1(t)g_1} e^{\overline{h}_2(t)g_2} e^{\overline{h}_3(t)g_3} e^{\overline{h}_4(t)g_4} e^{\overline{h}_5(t)g_5} e^{\overline{h}_6(t)g_6} \end{aligned} \tag{4.130}$$

The map $S \rightarrow (h_1, \ldots, h_6)$ and $S \rightarrow (\overline{h}_1, \ldots, \overline{h}_6)$ establish global diffeomorphism between Lie groups and $\Re^6$. h_i are the backward Philip Hall coordinates of S and

$\overline{h}_i$ are the forward Philip Hall coordinates of S. In addition, $S(t)$ satisfies the differential equation

$$\dot{S}(t) = S(t)\,(g_1u_1 + g_2u_2 + g_3u_3 + g_4u_4 + g_5u_5 + g_6u_6) \qquad S(0) = I$$

Differentiating yields to

$$\begin{pmatrix} \dot{h_1} \\ \dot{h_2} \\ \dot{h_3} \\ \dot{h_4} \\ \dot{h_5} \\ \dot{h_6} \end{pmatrix} = \begin{pmatrix} u_1 \\ u_2 \\ u_3 \\ u_4 + h_1u_1 \\ u_5 + h_2u_2 \\ u_6 + h_3u_3 \end{pmatrix} \qquad h_i(0) = 0 \tag{4.131}$$

The forward Philip Hall coordinates can be found out

$$\begin{pmatrix} \overline{h_1} \\ \overline{h_2} \\ \overline{h_3} \\ \overline{h_4} \\ \overline{h_5} \\ \overline{h_6} \end{pmatrix} = \begin{pmatrix} h_1 \\ h_2 \\ h_3 \\ h_4 + h_1 \times h_2 \\ h_5 + h_2 \times h_3 \\ h_6 + h_2 \times h_3 \end{pmatrix} \tag{4.132}$$

Now, the problem of finding the initial control U is tackled out.

4.5.3.2 Finding U

Now, the following initial value problem is considered:

$$\dot{S}^*(t) = S^*(t)\,(g_1u_1 + g_2u_2 + g_3u_3) \qquad S^*(0) = I \tag{4.133}$$

The backward Philip Hall coordinates h_i being known, the control $u(t)$ must be found that produces $S^*(t)$ having these coordinates. First, the forward coordinates can be obtained from Eq. (4.133). Then, the question is turned into finding a control for

$$S^*(t) = e^{\overline{h}_1(t)g_1} e^{\overline{h}_2(t)g_2} e^{\overline{h}_3(t)g_3} e^{\overline{h}_4(t)g_4} e^{\overline{h}_5(t)g_5} e^{\overline{h}_6(t)g_6}$$

Each exponential factor must be solved separately, then the results must be concatenated.

Remark 4.6 The symbol $\otimes$ is used for concatenation, for example $A \otimes B$ means A followed by B.

Since the first flow $e^{\overline{h}_1(t)g_1}$ is just along g_1, so the control input $u_1 = \overline{h}_1$ for the first control sequence. Similarly, for the second and third control sequences, the result input $u_2 = \overline{h}_2$ will result in the flow $e^{\overline{h}_2(t)g_2}$ and $u_3 = \overline{h}_3$ will result in the flow $e^{\overline{h}_3(t)g_3}$. So the control sequences $\overline{h}_1(t)g_1 \otimes \overline{h}_2(t)g_2 \otimes \overline{h}_3(t)g_3$ generate the flow $e^{\overline{h}_1(t)g_1}e^{\overline{h}_2(t)g_2}e^{\overline{h}_3(t)g_3}$.

Now that the flow along the Lie bracket direction $e^{\overline{h}_4(t)g_4} = e^{\overline{h}_4(t)[f,g_1]}$ is needed. Using the Campbell Baker Hausdorff formula, the following six control sequence (assuming $\overline{h}_4 > 0$),

$$\sqrt{\overline{h}_4(t)g_1} \otimes \sqrt{\overline{h}_4(t)g_2} \otimes \sqrt{\overline{h}_4(t)g_3}$$
$$\otimes \left(-\sqrt{\overline{h}_4(t)g_3}\right) \otimes \left(-\sqrt{\overline{h}_4(t)g_2}\right) \otimes \left(-\sqrt{\overline{h}_4(t)g_1}\right),$$

give rise to $e^{\overline{h}_4(t)g_4}$ idem for $e^{\overline{h}_5(t)g_5} = e^{\overline{h}_5(t)[f,g_2]}$ with

$$\sqrt{\overline{h}_5(t)g_1} \otimes \sqrt{\overline{h}_5(t)g_2} \otimes \sqrt{\overline{h}_5(t)g_3} \otimes \left(-\sqrt{\overline{h}_5(t)g_3}\right)$$
$$\otimes \left(-\sqrt{\overline{h}_5(t)g_2}\right) \otimes \left(-\sqrt{\overline{h}_5(t)g_1}\right)$$

give rise to $e^{\overline{h}_4(t)g_4}e^{\overline{h}_5(t)g_5}$ idem for $e^{\overline{h}_6(t)g_6} = e^{\overline{h}_6(t)[f,g_3]}$ with

$$\sqrt{\overline{h}_6(t)g_1} \otimes \sqrt{\overline{h}_6(t)g_2} \otimes \sqrt{\overline{h}_6(t)g_3} \otimes \left(-\sqrt{\overline{h}_6(t)g_3}\right)$$
$$\otimes \left(-\sqrt{\overline{h}_6(t)g_2}\right) \otimes \left(-\sqrt{\overline{h}_6(t)g_1}\right)$$

give rise to $e^{\overline{h}_4(t)g_4}e^{\overline{h}_5(t)g_5}e^{\overline{h}_6(t)g_6}$ so far a control U made of $1 \times 1 \times 1 \times 6 \times 6 \times 6 =$ 216 pieces have been obtained so $S(t)$ can be realized by 216 moves.

Remark 4.7 if $\overline{h}_4 < 0$ or $\overline{h}_5 < 0$ or $\overline{h}_6 < 0$, just interchange respectively f and g_1, f and g_2, f and g_3 [130].

4.5.4 Determination of the Minimum Energy Trajectories

The problem may be formulated as:

$$\min \frac{1}{2}\int_0^1 \left(u^T(t)u(t)dt\right)$$
$$\text{subject to} \quad \dot{X} = f(X) + \sum_{i=1}^{m} g_i(X)U_i$$

In this paragraph, the least squares optimal control problem for an affine system from drift from X_0 to X_f in one unit time. The steering problem is assumed to have a solution (by Chow's theorem, this is guaranteed when the controllability Lie algebra generated by $g_1, \dots, g_m$ is of rank n for all X) [52].

A heuristic derivation is given from the calculus of variations of the necessary conditions for optimality. Constraints are incorporated into the cost function using a Lagrange multiplier function $\lambda(t) \in \Re^n$ to get

$$J(X, \lambda, U) = \int_0^1 \left(\frac{1}{2} u^T(t) u(t) - \lambda^T \left(\dot{X} - f(X) - \sum_{i=1}^m g_i(X) U_i \right) \right) dt \quad (4.134)$$

The Hamiltonian function is introduced as:

$$H(X, \lambda, U) = \frac{1}{2} u^T(t) u(t) + \lambda^T \left(f(X) + \sum_{i=1}^m g_i(X) U_i \right) \quad (4.135)$$

This yields to

$$J(X, \lambda, U) = -\lambda^T(t) X(t)|_0^1 + \int_0^1 \left(H(X, \lambda, U) + \dot{\lambda}^T X \right) dt \quad (4.136)$$

Consider the variations in J caused by variations in the control input U

$$\delta J(X, \lambda, U) = -\lambda^T(t) \delta X(t)|_0^1 + \int_0^1 \left(\frac{\partial H}{\partial X} \delta X + \frac{\partial H}{\partial U} \delta U + \dot{\lambda}^T \delta X \right) dt \quad (4.137)$$

If the optimal input has been found, a necessary condition for stationarity is that the first variation above be zero for all variations δU and δX.

$$\begin{aligned} \dot{\lambda} &= -\frac{\partial H}{\partial X} \\ 0 &= \frac{\partial H}{\partial U} \end{aligned} \quad (4.138)$$

From the second of these equations, it follows that the optimal inputs are given by:

$$U_i = -\lambda_i^T g_i(X)$$

The optimal inputs satisfy the differential equations

$$\dot{U} = \Omega(\lambda, X) U + \begin{pmatrix} \lambda^T [f, g_1] \\ \cdots \\ \lambda^T [f, g_m] \end{pmatrix} \quad (4.139)$$

with

$$\Omega(\lambda, X) U = \begin{pmatrix} 0 & \lambda^T [g_1, g_2] & \lambda^T [g_1, g_3] \\ -\lambda^T [g_1, g_2] & 0 & \lambda^T [g_2, g_3] \\ -\lambda^T [g_1, g_3] & \lambda^T [g_2, g_3] & 0 \end{pmatrix}$$

4.5.5 Determination of Time Optimal Trajectories

The subject of this section is to formulate the trajectory generation problem in minimum time as this system has bounds on the magnitudes of the inputs and the states. The velocity is assumed to be linearly variable. As the set of allowable inputs is convex, the time optimal paths result from saturating the inputs at all times (or zero for singular control). For a linear time-invariant controllable system with bounded control inputs, the time-optimal control solution to a typical two point boundary value problem is a bang-bang function with a finite number of switches. Time optimal trajectory generation can be formulated as follows [25]:

$$\min \int_0^T dt \tag{4.140}$$

subject to

$$\begin{aligned} \dot{x} &= V \sin\chi \cos\gamma + W_x \\ \dot{y} &= V \cos\chi \cos\gamma + W_y \\ \dot{z} &= V \sin\gamma + W_z \end{aligned} \tag{4.141}$$

$$\begin{aligned} \dot{V} &= u_1 \\ \dot{\chi} &= u_2 \\ \dot{\gamma} &= u_3 \end{aligned} \tag{4.142}$$

initial and final conditions

$$\begin{aligned} &x(0) = x_0, && y(0) = y_0, && z(0) = z_0, \\ &\chi(0) = \chi_0, && \gamma(0) = \gamma_0, && V(0) = V_0 \\ &x(T) = x_f, && y(T) = y_f, && z(T) = z_f, \\ &\chi(T) = \chi_f, && \gamma(T) = \gamma_f, && V(T) = V_f \end{aligned} \tag{4.143}$$

Limitations on the control inputs

$$|u_1| \le u_{1max} \qquad |u_2| \le u_{2max} \qquad |u_3| \le u_{3max} \tag{4.144}$$

and on state variables

$$|V| \le V_{max} \qquad |\gamma| \le \gamma_{max} \tag{4.145}$$

Remark 4.8 This formulation is a generalization of Zermelo's navigation problem, where the problem consists of finding the quickest nautical path for a ship at sea in the presence of currents.

For points that are reachable, the resolution is based on the Pontryagin Minimum Principle which constitutes a generalization of Lagrange problem of the calculus of variations. It is a local reasoning based on the comparison of trajectories corresponding to infinitesimally close control laws [192]. It provides necessary conditions for paths to be optimal. Of course, the kinematic model used below implies a perfect response to the turn commands. A major reason for using the kinematic model is the fact that only necessary conditions for optimality exist for the second order model (given by Pontryagin minimum principle) [93, 95, 224].

The Hamiltonian, formed by adjoining the state equation with the appropriate adjoint variable $\lambda_1, \dots, \lambda_6$, is classically defined as follows:

$$\begin{aligned} H &= 1 + \lambda_1(V \sin\chi \cos\gamma + W_x) + \lambda_2(V \cos\chi \cos\gamma + W_y) \\ &\quad + \lambda_3(V \sin\gamma + W_z) + \lambda_4 u_1 + \lambda_5 u_2 + \lambda_6 u_3 \end{aligned} \tag{4.146}$$

where λ represents the Lagrange multiplier. The optimal control input must satisfy the following set of necessary conditions

$$\dot{X} = \frac{\partial H}{\partial \lambda} \quad \text{where X(0), X(T) are specified} \tag{4.147}$$

$$\dot{\lambda} = -\frac{\partial H}{\partial X} \quad \text{where } \lambda(0), \lambda(T) \text{ are free} \tag{4.148}$$

With the transversality condition

$$H(T) = 0 \tag{4.149}$$

The co-state variables are free, i.e. unspecified, at both the initial and final times because the corresponding state variables of the system are specified. A first interesting result is the determination of a sufficient family of trajectories, i.e. a family of trajectories containing an optimal solution for linking any two configurations.

As the drift is not an explicit function of time t, a first integral of the two point boundary value problem exists and thus the Hamiltonian H is constant on the optimal trajectory. Because $H(T) = 0$ from the transversality condition,

$$H(t) = 0, \quad \forall t \in [0, t_f]$$

The co-state equations are then obtained in the standard fashion by differentiating the negative of the Hamiltonian with respect to the states.

Lagrange Multipliers Analysis The first order necessary conditions must be satisfied by formulating the differential equations for the costates. The adjoint equations are the first part of the necessary conditions (4.148) where

$$\begin{aligned}
\dot{\lambda_1} &= 0 \\
\dot{\lambda_2} &= 0 \\
\dot{\lambda_3} &= 0 \\
\dot{\lambda_4} &= \lambda_1 \cos\chi \cos\gamma - \lambda_2 \sin\chi \cos\gamma + \lambda_3 \sin\gamma \\
\dot{\lambda_5} &= \lambda_1 V \sin\chi \cos\gamma - \lambda_2 V \cos\chi \cos\gamma = \lambda_1(\dot{y} - W_y) - \lambda_2(\dot{x} - W_x) \\
\dot{\lambda_6} &= \lambda_1 V \cos\chi \sin\gamma + \lambda_2 V \sin\chi \sin\gamma - \lambda_3 V \cos\gamma
\end{aligned} \tag{4.150}$$

Integrating the first three multiplier dynamics, the following relations are obtained

$$\begin{aligned}
\lambda_1 &= \mu \sin\zeta = \text{constant} \\
\lambda_2 &= \mu \cos\zeta = \text{constant} \\
\lambda_3 &= \text{constant} \\
\dot{\lambda}_4 &= -\frac{\mu}{2}\cos(\chi + \gamma - \zeta) - \frac{\mu}{2}\cos(\chi - \gamma - \zeta) + \lambda_3 \sin\gamma \\
\dot{\lambda}_5 &= -\frac{\mu V}{2}\sin(\chi + \gamma - \zeta) - \frac{\mu V}{2}\sin(\chi - \gamma - \zeta) \\
\dot{\lambda}_6 &= \frac{\mu V}{2}\sin(\chi + \gamma - \zeta) + \frac{\mu V}{2}\cos(\chi - \gamma + \zeta) - \lambda_3 V \sin\gamma
\end{aligned} \tag{4.151}$$

$$\lambda_5 = \lambda_1 y - \lambda_2 x - \lambda_1 W_y t + \lambda_2 W_x t + \lambda_{50} \tag{4.152}$$

Defining the Hamiltonian and multiplier dynamics in this way, the minimum principle of Pontryagin states that the control variable must be chosen to minimize the Hamiltonian at every instant.

$$H(X, u^*) \leq H(X, u) \tag{4.153}$$

On the optimal trajectory, the optimal control u^* must satisfy:

$$\lambda_4 u_1^* + \lambda_5 u_2^* + \lambda_6 u_3^* \leq \lambda_4 u_1 + \lambda_5 u_2 + \lambda_6 u_3 \tag{4.154}$$

Minimization of the Hamiltonian function subject to the control constraints requires that

$$u_j = -\operatorname{sgn}\{S_j\} = -\operatorname{sgn}\left\{\sum_{i=1}^{n} B_{ij}\lambda_i,\ j = 1\dots m\right\} \tag{4.155}$$

where S_j is the switching function associated with the jth control input u_j and the signum function is defined as

$$\operatorname{sgn}\{S_j\} = \begin{cases} +1 & \text{if } S_j > 0 \\ -1 & \text{if } S_j < 0 \end{cases} \tag{4.156}$$

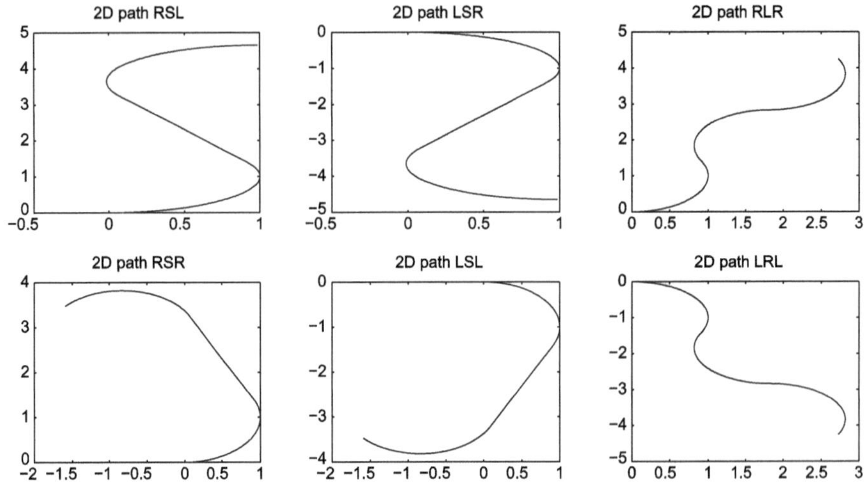

Fig. 4.4 The six sequences of Dubins

and it becomes singular for

$$-1 < u_j < +1 \quad \text{if } S_j = 0$$

Leading to the following solution

$$\begin{aligned} u_1^* &= -u_{1max}\,\text{sign}(\lambda_4) \\ u_2^* &= -u_{2max}\,\text{sign}(\lambda_5) \\ u_3^* &= -u_{3max}\,\text{sign}(\lambda_6) \end{aligned} \tag{4.157}$$

Thus, the following multipliers can be integrated as

$$\begin{aligned} \lambda_4 &= -\frac{\mu}{2}\frac{\sin(\chi+\gamma-\zeta)}{\delta_2 u_{2max}+\delta_3 u_{3max}} - \frac{\mu}{2}\frac{\sin(\chi-\gamma-\zeta)}{\delta_2 u_{2max}-\delta_3 u_{3max}} + \frac{\lambda_3\cos\gamma}{\delta_3 u_{3max}} \\ \lambda_5 &= \mu\sin\zeta\,\left(y - W_y t\right) - \mu\cos\zeta\,(x - W_x t) \end{aligned} \tag{4.158}$$

In 2D, according to Dubins' theorem [58] a mobile robot minimum time optimal trajectory under maximum control constraint and constant velocity has six solutions $\{RSL, RSR, LSL, LSR, RLR, LRL\}$ where R represents Right, S: straight and L Left. Such paths are a concatenation of an arc of a minimum-radius circle (either in the positive or negative direction) with either an arc of a minimum radius circle (in the opposite direction) or with a straight segment. Figure 4.4 shows these six curves.

4.5.5.1 Singular Controls

Let u_k denote a component of the control vector U. A regular control u_k is a control where the optimality condition $H_{u_k} = 0$ explicitly contains the control u_k, so that

$H_{u_k}u_k$ is not identically zero. The Weierstrass condition is a necessary condition for a regular optimal control to be a strong relative minimum.

A singular control u_k occurs when $H_{u_k}u_k \equiv 0$. It is an important case as this Hamiltonian contains control variables linearly. Such paths are a concatenation of an arc of a minimum-radius circle (either in the positive or negative direction) with either an arc of a minimum radius circle (in the opposite direction) or with a straight segment.

The generalized Legendre-Clebsch condition

$$(-1)^{\iota}\frac{\partial}{\partial U}\left(\frac{d^2 H_u}{dt^2}\right) \geq 0 \tag{4.159}$$

where ι denotes the order of the singular control. The Weierstrass condition and the Legendre-Clebsch condition can be applied to individual or combinations of regular controls and the generalized Legendre-Clebsch condition can be applied to individual singular controls.

1. For the multiplier λ_4, when the singular control occurs $V = V_{max}$ and $\lambda_4 = \dot{\lambda}_4 = 0$ thus, the relation

$$\dot{\lambda}_4 V = -\lambda_1 (\dot{x} - W_x) - \lambda_2 (\dot{y} - W_y) + \lambda_3 (\dot{z} - W_z)$$

Integrating for the singular control, the following relation is obtained:

$$\lambda_1 x - \lambda_2 y + \lambda_3 z = \lambda_1 W_x - \lambda_2 W_y + \lambda_3 W_z + \lambda_7$$

where λ_7 is an integration constant.

2. Because the values of $\lambda_1, \lambda_2, \lambda_{50}$ are constant, each value of λ_5 defines a line parallel to the characteristic direction, if $W_x = W_y = 0$. The line defined by $\lambda_5 = 0$ is the line on some switching and straight line travel must occur. Straight lines and changing in turning direction on the optimal path must occur on a single line.

$$y = \frac{\lambda_2}{\lambda_1}x + W_y t - \frac{\lambda_2}{\lambda_1}W_x t - \frac{\lambda_{50}}{\lambda_1} \tag{4.160}$$

3. If both following relations are compared for the singular control in λ_6:

$$0 = \dot{\lambda}_6 = \lambda_1 V \cos\chi \sin\gamma + \lambda_2 V \sin\chi \sin\gamma - \lambda_3 V \cos\gamma$$

and the nonholonomic constraint

$$0 = \dot{x}\cos\chi \sin\gamma + \dot{y}\sin\chi \sin\gamma - \dot{z}\cos\gamma$$

thus

$$\begin{aligned} \dot{x} &= \lambda_1 V \\ \dot{y} &= \lambda_2 V \\ \dot{z} &= \lambda_3 V \end{aligned} \tag{4.161}$$

Four cases can be derived from the previous relations.

- $\lambda_1 = \lambda_2 = \lambda_3 = 0$: The optimal trajectory is represented by a point.
- $\lambda_1 = \lambda_2 = 0$: The optimal trajectory is represented by a straight line along the Up axis.
- $\lambda_1 = \lambda_3 = 0$: The optimal trajectory is represented by a straight line along the North axis.
- $\lambda_2 = \lambda_3 = 0$: The optimal trajectory is represented by a straight line along the East axis.

It is necessary to fuse a boundary-valued optimal control with a singular value.

4.5.5.2 Paths

The number of switches in the controls u_1, u_2, u_3 depend on the corresponding functions $\lambda_4, \lambda_5, \lambda_6$ respectively. When the initial and final points are more than two turn radii apart, the minimum time path comprises 3 segments. All straight lines and changing in turning direction on the optimal path must occur on a single line. These solutions can be written in the following concise form:

$$u_1^* \in \left\{ \begin{pmatrix} \delta_1 u_{1max} \\ 0 \\ \pm\delta_1 u_{1max} \end{pmatrix}, \begin{pmatrix} \delta_1 u_{1max} \\ -\delta_1 u_{max} \\ \delta_1 u_{1max} \end{pmatrix} \right\} \tag{4.162}$$

$$u_2^* \in \left\{ \begin{pmatrix} \delta_2 u_{2max} \\ 0 \\ \pm\delta_2 u_{2max} \end{pmatrix}, \begin{pmatrix} \delta_2 u_{2max} \\ -\delta_2 u_{2max} \\ \delta_2 u_{2max} \end{pmatrix} \right\} \tag{4.163}$$

$$u_3^* \in \left\{ \begin{pmatrix} \delta_3 u_{3max} \\ 0 \\ \pm\delta_3 u_{3max} \end{pmatrix}, \begin{pmatrix} \delta_3 u_{3max} \\ -\delta_3 u_{3max} \\ \delta_3 u_{3max} \end{pmatrix} \right\} \tag{4.164}$$

with

$$\delta_1 = -\operatorname{sign}(\lambda_4) = \pm 1 \tag{4.165}$$

$$\delta_2 = -\operatorname{sign}(\lambda_5) = \pm 1 \tag{4.166}$$

$$\delta_3 = -\operatorname{sign}(\lambda_6) = \pm 1 \tag{4.167}$$

Analysis of the First Set of Solutions S1 In this paragraph, emphasis is put on an analytical resolution of the first set of time optimal control problem solutions.

S1:

$$u_1^* = \begin{pmatrix} \delta_1 u_{1max} \text{ for } 0 \le t \le t_1 \\ -\delta_1 u_{1max} \text{ for } t_1 \le t \le t_2 \\ \delta_1 u_{1max} \text{ for } t_2 \le t \le T \end{pmatrix} \tag{4.168}$$

$$u_2^* = \begin{pmatrix} \delta_2 u_{2max} \text{ for } 0 \le t \le t_1' \\ -\delta_2 u_{2max} \text{ for } t_1' \le t \le t_2' \\ \delta_2 u_{2max} \text{ for } t_2' \le t \le T \end{pmatrix} \tag{4.169}$$

$$u_3^* = \begin{pmatrix} \delta_3 u_{3max} \text{ for } 0 \le t \le t_1'' \\ -\delta_3 u_{3max} \text{ for } t_1'' \le t \le t_2'' \\ \delta_3 u_{3max} \text{ for } t_2'' \le t \le T \end{pmatrix} \tag{4.170}$$

Taking into account the following conditions

$$\begin{array}{cccc} T > 0 & t_1 \ge 0 & t_1 \le t_2 & t_2 \le T \\ & t_1' \ge 0 & t_1' \le t_2' & t_2' \le T \\ & t_1'' \ge 0 & t_1'' \le t_2' & t_2'' \le T \end{array} \tag{4.171}$$

By integration of Eq. (4.168), one can obtain;

$$V^*(t) = \begin{pmatrix} \delta_1 u_{1max} t + V_0 \text{ for } 0 \le t \le t_1' \\ -\delta_1 u_{1max} t + V_1 \text{ for } t_1' \le t \le t_2' \\ \delta_1 u_{1max} t + V_2 \text{ for } t_2' \le t \le T \end{pmatrix} \tag{4.172}$$

$$\chi^*(t) = \begin{pmatrix} \delta_2 u_{2max} t + \chi_0 \text{ for } 0 \le t \le t_1 \\ -\delta_2 u_{2max} t + \chi_1 \text{ for } t_1 \le t \le t_2 \\ \delta_2 u_{2max} t + \chi_2 \text{ for } t_2 \le t \le T \end{pmatrix} \tag{4.173}$$

$$\gamma^*(t) = \begin{pmatrix} \delta_3 u_{3max} t + \gamma_0 \text{ for } 0 \le t \le t_1'' \\ -\delta_3 u_{3max} t + \gamma_1 \text{ for } t_1'' \le t \le t_2'' \\ \delta_3 u_{3max} t + \gamma_2 \text{ for } t_2'' \le t \le T \end{pmatrix} \tag{4.174}$$

The integration constants $\gamma_1, \gamma_2, \chi_1, \chi_2, V_1, V_2$ will be determined using the initial and final conditions as well as the continuity condition.

$$\begin{array}{ll} V_1 = 2\delta_1 u_{1max} t_1' + V_0 & V_2 = -2\delta_1 u_{1max} t_2' + V_1 \\ \chi_1 = 2\delta_2 u_{2max} t_1 + \chi_0 & \chi_2 = -2\delta_2 u_{2max} t_2 + \chi_1 \\ \gamma_1 = 2\delta_3 u_{3max} t_1' + \gamma_0 & \gamma_2 = -2\delta_3 u_{3max} t_2' + \gamma_1 \end{array} \tag{4.175}$$

The final conditions give

$$\begin{aligned} V_f &= V_2 + \delta_1 u_{1max} T \\ \chi_f &= \chi_2 + \delta_2 u_{2max} T \\ \gamma_f &= \gamma_2 + \delta_3 u_{3max} T \end{aligned} \tag{4.176}$$

With this relation, one obtains the first equation relying the transition times with the final time.

$$t_2 = \frac{1}{2}\left(T + 2t_1 - \frac{V_f - V_0}{\delta_1 u_{1max}}\right)$$

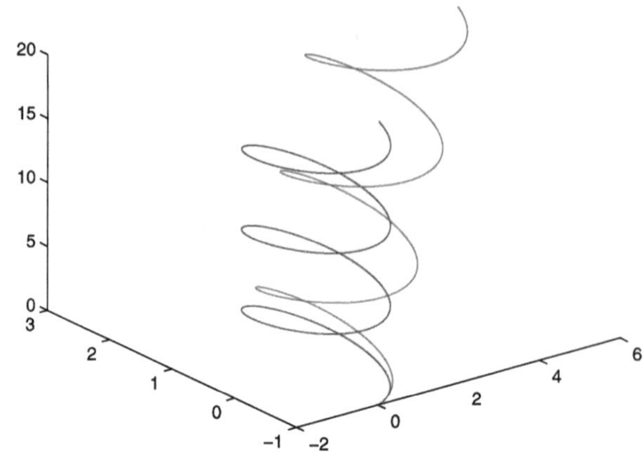

Fig. 4.5 (Color online) Trim helicoidal motion without (*blue*) and with (*red*) wind

$$t_2' = \frac{1}{2}\left(T + 2t_1' - \frac{\chi_f - \chi_0}{\delta_2 u_{2max}}\right) \tag{4.177}$$

$$t_2'' = \frac{1}{2}\left(T + 2t_1'' - \frac{\gamma_f - \gamma_0}{\delta_3 u_{3max}}\right)$$

Next, the vertical equation in z is integrated, followed by the integration of the horizontal equations in x, y, respecting the boundary values conditions and the continuity at the switching points. With the condition $H(t) = 0\ \forall t \in [0, T]$, transcendental equations are obtained.

T is the smallest positive solution of these equations to be solved numerically. As this family is small enough and sufficiently well specified, it is possible to compare the cost of candidate trajectories by means of a numerical technique. The existence of a candidate solution time T cannot be ensured in any condition, particularly when wind is too strong.

Simulation Results A well known situation concerns the trim trajectories in the air path. They are obtained for $t_1' = 0, t_2' = T$.

Figure 4.5 shows the effect of a steady wind when $W_x = 0.25V$ on the trim trajectory (in blue). The red curve represents a helix with a translation in the x direction. In 2D plane, a trochoid curve is obtained. The influence of the wind is bigger as time passes by. Figure 4.6 represents the trochoidal variation in the x-y plane (in red) and a trim circle (in blue). Figure 4.7 presents the general case where $t_1' \neq 0, t_1 \neq 0, t_2' \neq 0, t_2 \neq 0, T \neq 0$, with the initial conditions $x_0 = 0$; $y_0 = 0, z_0 = 0, \chi_0 = 0, \gamma_0 = 0$.

4.5.5.3 Numerical Approach

Optimal control problems are often solved numerically via direct methods. In recent years considerable attention has been focused on a class of direct transcription

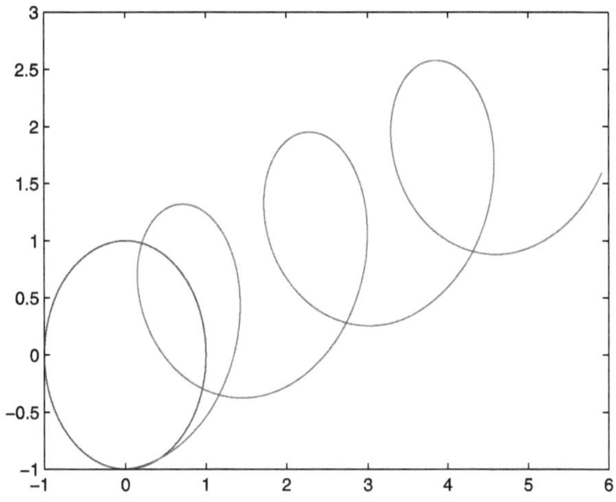

Fig. 4.6 (Color online) Trochoidal variation

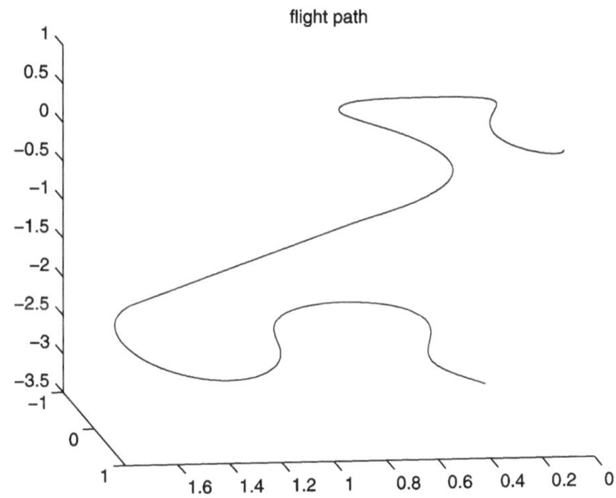

Fig. 4.7 3D general variation

method called pseudo-spectral or orthogonal collocation methods. In a pseudo spectral method, a finite basis of global interpolating polynomial is used to approximate the state at a set of discretization points. The time derivative of the state is approximated by differentiating the interpolating polynomial and constraining the derivative to be equal to the vector field at a finite set of collocation points. Although any set of unique collocation points can be chosen, an orthogonal collocation is chosen, i.e. the collocation point are the roots of an orthogonal polynomial (or linear combinations of such polynomials and their derivatives). Pseudo-spectral methods are commonly implemented via orthogonal collocation.

Within the class of pseudo-spectral methods, there are two different implementation strategies: local and global approaches. In a local approach, the time interval is divided into a large number of subintervals called segments or finite elements and a small number of collocation points are used within each segment. The segments are then linked via continuity condition on the state, the independent variable and possibly the control. The rationale for using local collocation is that the discretization points are located so that they support the local behavior of the dynamics.

In global collocation form, a single segment is used across the entire interval and the number of collocation points is varied. For smooth problems, global collocation is more accurate than local collocation for a given number of total collocation points. For non smooth problems, the accuracies of global and local collocation methods are comparable.

To improve the accuracy of the direct optimization solutions and to enlarge the convergence domain of the indirect methods, a hybrid approach is proposed to solve the optimal control problem. This cascaded computational scheme has become widely applied. The key idea is to extract the co-states and other control structure information from a nonlinear programming approach as a first step. The indirect shooting method is then used to refine the solutions.

The three major steps to solve for the optimal maneuver solutions and to validate the results based on the first order optimality conditions.

1. The kinematic and dynamic differentiation equations are discretized using the Euler or trapezoidal method. Commercially available software is used to get the preliminary and approximate control structures, switching times and initial co-states
2. Using the results from step 1 as the initial guess, this software is used as a shooting method to solve the two point boundary value problem. The constraints include the final time conditions and the invariance of the Hamiltonian.
3. The results from step 2, together with the originally known initial time state conditions, are used to solve for the dynamic system response by integrating the kinematic and dynamic equations forward in time. The Hamiltonian history and the final state errors are the validation criteria.

This approach can only guarantee that the found solutions are local extrema.

4.6 Parametric Curves

To lead the lighter than air robot from an initial configuration $q(t_i) = q_i$ to a final configuration $q(t_f) = q_f$ in the absence of obstacles, a trajectory $q(t)$ for $t \in [t_i, t_f]$ has to be planned. The trajectory $q(t)$ can be broken down into a geometric path $q(s)$ with $\frac{dq(s)}{ds} \neq 0$ for any value of s and a timing law $s = s(t)$ with the parameter s varying between $s(t_i) = s_i$ and $s(t_f) = s_f$ in a monotonic fashion, i.e. with $\dot{s}(t) \geq 0$ for $t \in [t_i, t_f]$. A possible choice for s is the arc length along the path; in this case it would be $s_i = 0$ and $s_f = L$ where L is the length of the path. The above space time separation implies that

$$\dot{q} = \frac{dq}{dt} = \frac{dq}{ds}\dot{s} = q'\dot{s}$$

where the prime symbol denotes differentiation with respect to s. The generalized velocity vector is then obtained as the product of the vector q', which is directed as the tangent to the path in configuration space, by the scalar $\dot{s}$ that varies its modulus. Nonholonomic constraints can then be rewritten as

$$A(q)\dot{q} = A(q)q'\dot{s} = 0$$

if $\dot{s} \neq 0$ for $t \in [t_i, t_f]$, one has

$$A(q)q' = 0$$

This condition that must be verified at all points by the tangent vector on the configuration space path, characterizes the motion of geometric path admissibility induced by the kinematic constraint that actually affects generalized velocities.

Path planning focuses on finding a path through free space from an initial to a final location. The focus in this section in on turning a sequence of configurations into a smooth curve that is then passed to the control system of the vehicle. In 2D the curves fall into two categories

- Curves whose coordinates have a closed form expressions for example B-splines, quintic polynomials or polar splines
- Curves whose curvature is a function of their arc length for example clothoids, cubic spirals, quintic G^2 splines or intrinsic splines

For non holonomic vehicles such as mobile robots or aerial vehicles, dynamic model and actuators constraints that directly affect path are used to reject infeasible paths. The term feasible means that the path will be continuously flyable and safe. The flyable path should be smooth, i.e without twists and cusps. The smoothness of the path is determined by amount of bending of the path, measured by curvature and torsion of the path [4, 6].

The curvilinear abscissa s being considered instead of the time, the curve C(s) represents the motion of this vehicle in R^3, where

$$V = \frac{ds}{dt} \tag{4.178}$$

The shape of a space curve can be completely captured by its curvature and torsion. The tangent, normal and binormal of the curve $C(s)$, in the Frenet Serret frame, are defined as:

$$T(s) = \frac{C'(s)}{\|C'(s)\|} \qquad N(s) = \frac{C'(s) \times C''(s)}{\|C'(s) \times C''(s)\|} \qquad b(s) = T(s) \times N(s) \tag{4.179}$$

This tangent is constrained to have unity norm ($'$ represents derivation versus the curvilinear abscissa s). $\kappa(s)$ is called the curvature of the motion:

$$\kappa(s) = \frac{C'(s) \times C''(s)}{\|C'(s)\|^3} \tag{4.180}$$

and the torsion of the motion is

$$\tau(s) = \frac{(C'(s) \times C''(s))C'''}{\|C'(s) \times C''(s)\|^2} \tag{4.181}$$

If a non vanishing curvature and a torsion are given as smooth functions of s, theoretically both equations can be integrated to find the numerical values of the corresponding space curve (up to a rigid motion). The shape of a 3D curve can be completely captured by its curvature and torsion functions. Hence, they are considered to be a set of intrinsic and complete shape features of the curve $C(s)$ and expressed by the differential system.

$$\begin{pmatrix} \dot{T}(s) \\ \dot{N}(s) \\ \dot{b}(s) \end{pmatrix} = \begin{pmatrix} 0 & \kappa & 0 \\ -\kappa & 0 & \tau \\ 0 & -\tau & 0 \end{pmatrix} \begin{pmatrix} T(s) \\ N(s) \\ b(s) \end{pmatrix}$$

The length of the Darboux vector, also called total curvature, includes both of the above features:

$$\Theta = \tau(s)T(s) + \kappa(s)B(s) \tag{4.182}$$

Θ indicates how the entire frame rotates, making it the measure of the structural variation in C. The entire frame rotate about Θ at the angular rate of $\|\Theta\|$.

Thus, the differential equation describing the evolution of the Frenet formula for parameterized curves in $\Re^3$ for a unit speed curve can be written as:

$$\frac{d\Phi(s)}{ds} = \Phi(s)Sk \begin{pmatrix} \tau \\ 0 \\ \kappa \end{pmatrix} \tag{4.183}$$

If a non vanishing curvature and torsion are given as smooth functions of s, theoretically Eq. (4.183) when integrated can give the numerical value of the corresponding space curve, up to a rigid motion.

The path planning generates a feasible flight path for an aerial vehicle to reach the target. Geometrically, admissible paths for the lighter than air robot are the solution of this system. In 3D space, the following flight path is characterized by:

$$\begin{aligned} dx &= \sin\chi \cos\gamma ds \\ dy &= \cos\chi \cos\gamma ds \\ dz &= \sin\gamma ds \end{aligned} \tag{4.184}$$

Two non-holonomic constraints can thus be deduced:

$$\begin{aligned} dx\cos\chi - dy\sin\chi &= 0 \\ \{dx\sin\chi + dy\cos\chi\}\sin\gamma - dz\cos\gamma &= 0 \end{aligned} \tag{4.185}$$

Using the Frenet-Serret formulation,curvature κ can be deduced:

$$\kappa(s) = (\gamma'^2 + \chi'^2\cos^2\gamma)^{-1/2} \tag{4.186}$$

as well as torsion τ

$$\tau(s) = (\chi'\gamma''\cos\gamma + 2\chi'\gamma'^2\sin\gamma - \gamma'\chi''\cos\gamma - \gamma'\chi'^2\cos\chi\cos\gamma\sin^2\gamma\sin\chi + \chi'^3\sin\gamma\cos^2\gamma) / (\gamma'^2 + \chi'^2\cos^2\gamma) \tag{4.187}$$

Remark 4.9 The Frenet frame equations are pathological for example when the curve is perfectly straight or when the curvature vanishes momentarily.

The purpose of the following sections is to propose a 3D flight path to the aerial vehicle joining the initial and final configurations.

The inputs of this path planning algorithm are

- the initial configuration: $x_i, y_i, z_i, \gamma_i, \chi_i, V_i$
- the final configuration: $x_f, y_f, z_f, \gamma_f, \chi_f, V_f$.

Depending on these parameters, many possibilities exist. The most obvious one is to propose a polynomial variation of x, y and z.

4.6.1 Cartesian Polynomials

This problem can be solved by interpolating via Cartesian polynomials using the following cubic polynomials versus a normalized arclength $0 \le s \le 1$

$$\begin{aligned} x(s) &= s^3 x_f - (s-1)^3 x_i + \alpha_x s^2(s-1) + \beta_x s(s-1)^2 \\ y(s) &= s^3 y_f - (s-1)^3 y_i + \alpha_y s^2(s-1) + \beta_y s(s-1)^2 \\ z(s) &= s^3 z_f - (s-1)^3 z_i + \alpha_z s^2(s-1) + \beta_z s(s-1)^2 \end{aligned} \tag{4.188}$$

that automatically satisfy the boundary conditions on x, y, z. The orientation at each point being related to x′, y′, z′, it is also necessary to impose the additional boundary conditions

$$\begin{aligned} x'(0) &= \cos\gamma_i \sin\chi_i \\ y'(0) &= \cos\gamma_i \cos\chi_i \\ z'(0) &= \sin\gamma_i \end{aligned} \tag{4.189}$$

and

$$\begin{aligned} x'(1) &= \cos\gamma_f \sin\chi_f \\ y'(1) &= \cos\gamma_f \cos\chi_f \\ z'(1) &= \sin\gamma_f \end{aligned} \tag{4.190}$$

Deriving once the equations (4.188), the following relations are obtained:

$$\begin{aligned} x'(s) &= 3s^2 x_f - 3(s-1)^2 x_i + \alpha_x s(3s-2) + \beta_x (3s-1)(s-1) \\ y'(s) &= 3s^2 y_f - 3(s-1)^2 y_i + \alpha_y s(3s-2) + \beta_y (3s-1)(s-1) \\ z'(s) &= 3s^2 z_f - 3(s-1)^2 z_i + \alpha_z s(3s-2) + \beta_z (3s-1)(s-1) \end{aligned} \tag{4.191}$$

and deriving twice the equations (4.188),

$$\begin{aligned} x''(s) &= 6s x_f - 6(s-1) x_i + \alpha_x (6s-2) + \beta_x (6s-2) \\ y''(s) &= 6s y_f - 6(s-1) y_i + \alpha_y (3s-2) + \beta_y (6s-2) \\ z''(s) &= 6s z_f - 6(s-1) z_i + \alpha_z (6s-2) + \beta_z (6s-2) \end{aligned} \tag{4.192}$$

The geometric inputs that drive the robot along the Cartesian path are

$$\begin{aligned} V(s) &= \pm\sqrt{(x'(s))^2 + (y'(s))^2 + (z'(s))^2} \\ V &= \frac{1}{T}\sqrt{(x'(s))^2 + (y'(s))^2 + (z'(s))^2} \end{aligned} \tag{4.193}$$

Resolving for the various parameters, the following equalities are obtained

$$\begin{aligned} \alpha_x &= \cos\gamma_f \sin\chi_f - 3x_f \\ \alpha_y &= \cos\gamma_f \cos\chi_f - 3y_f \\ \alpha_z &= \sin\gamma_f - 3z_f \end{aligned} \tag{4.194}$$

and

$$\begin{aligned} \beta_x &= \cos\gamma_i \sin\chi_i + 3x_i \\ \beta_y &= \cos\gamma_i \cos\chi_i + 3y_i \\ \beta_z &= \sin\gamma_i + 3z_i \end{aligned} \tag{4.195}$$

The evolution of the vehicle orientation along the path can then be computed for the flight path angle

$$\gamma = \operatorname{asin}\left(\frac{dz/ds}{V(s)}\right) \tag{4.196}$$

and the heading angle

$$\chi = \operatorname{atan} 2\left(\frac{dx}{ds}, \frac{dy}{ds}\right) + k\pi; \quad k = 0, 1 \tag{4.197}$$

The two possible choices for k account for the fact that the same Cartesian path may be followed moving forward (k=0) or backward (k=1). If the initial orientation is assigned, only one of the choices for k is correct.

This approach can be easily generalizable to a fourth or fifth polynomial. However the main drawback is a complicated formulation of the curvature and the torsion making control of smoothness (twists and cusps) a difficult task. The approach followed in the following subsection aims to propose an easy formulation of these two parameters.

Fig. 4.8 Trim path

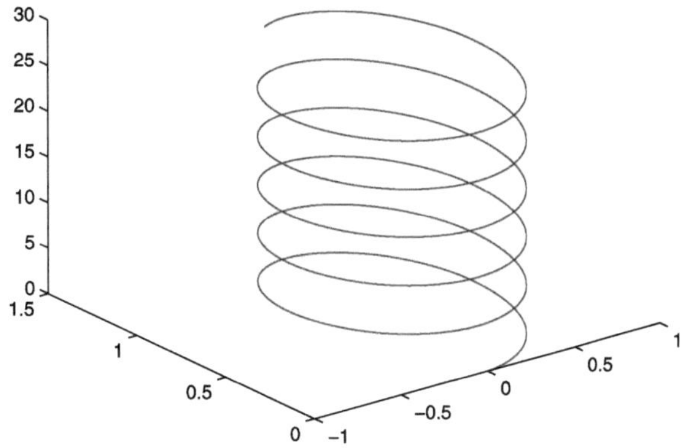

4.6.2 *Trim Flight Paths*

In this case, the flight path angle γ is constant while the angle χ is linearly varying versus s.

$$\chi(s) = \chi_0 + s\chi_1 \tag{4.198}$$

The parameters γ_0, χ_0, χ_1 being constants, the following relations can be proposed:

$$x(s) = x_i + \frac{\cos\gamma_0}{\chi_1}(\cos(\chi_0 + \chi_1 s) - \cos\chi_0) \tag{4.199}$$

$$y(s) = y_i - \frac{\cos\gamma_0}{\chi_1}(\sin(\chi_0 + \chi_1 s) - \sin\chi_0) \tag{4.200}$$

$$z(s) = s\sin(\gamma_0) + z_i \tag{4.201}$$

It is a well-known fact that trim trajectories are represented in general by helices, with particular cases: straight motion or circle arcs (see Fig. 4.8) (The units are in 1000 m).

This particular case occurs when $\gamma_i = \gamma_f$ and the following relationships are verified

$$\chi_1 = \sin(\gamma_i)\frac{\chi_f - \chi_i}{z_f - z_i} \tag{4.202}$$

with the constraint between the initial and final positions:

$$[\chi_1(x_f - x_i) + \cos\gamma_i \sin\chi_i]^2 + [\chi_1(y_f - y_i) - \cos\gamma_i \cos\chi_i]^2 = \cos^2\gamma_i \tag{4.203}$$

L the length of the path being given by

$$L = \frac{\chi_f - \chi_i}{\chi_1} \tag{4.204}$$

For this kind of curves, curvature and torsion are constant

$$\begin{aligned}\kappa(s) &= \chi_1 \cos(\gamma_0)\\ \tau(s) &= \chi_1 \sin(\gamma_0)\end{aligned} \tag{4.205}$$

The role of the trajectory generator is to generate a feasible time trajectory for the aerial vehicle. Once the path has been calculated in the Earth fixed frame, motion must be investigated using the dynamic model and reference trajectories determined taking into account actuators constraints (inequality constraints) and the under-actuation (equality constraints) of an aerial vehicle and limitations on curvature and torsion.

4.6.3 Non Trim Flight Paths

In this paragraph, non trim trajectories are studied where the flight path angle γ is assumed in the first instance constant, then a linear function of s. The heading angle is assumed to be a linear or a quadratic function of s.

Remark 4.10 Transition curves are used for blending in the plane that is to round corners or for smooth transition between straight lines and circular arcs, two circular arcs or two straight lines. The resulting curves should have position and curvature continuity as well as continuity of the unit tangent vector. Such continuity is usually referred to in the Computer Aided Geometric Design as G^2 Continuity. Higher order of continuity G^k will ensure smoother transition curve with direct physical meaning on moving the lighter than air robot along its path.

The issues of continuity of the curvature and its derivative have led to novel parametric curves such as polar polynomial, cubic spiral, sums of sine and cosine, Pythagorean Hodograph curve, parameter splines of higher order, uniform quartic B-splines, η^3 splines.

4.6.3.1 Non Trim Trajectories at Constant Altitude

The initial $P_i = (x_i, y_i, \chi_i)^T$ and final configurations $P_f = (x_f, y_f, \chi_f)^T$ are said symmetrical if $\zeta = \frac{\chi_i + \chi_f}{2}$ where $\varsigma = \arctan \frac{y_f - y_i}{x_f - x_i}$ the orientation of P_f versus P_i. Let's define the Euclidean distance $d = \sqrt{(x_f - x_i)^2 + (y_f - y_i)^2}$ and the deflection $\zeta = \chi_f - \chi_i$. A path is represented by the pair (ℓ, κ) where ℓ is a positive length and κ: $[-\frac{\ell}{2}, \frac{\ell}{2}]$. Let's suppose that $x(0) = y(0) = \chi(0) = 0$. Knowing the path curvature $\kappa(s)$ and initial conditions allow the construction of the path:

$$\begin{aligned} \chi(s) &= \int_0^s \kappa(u)du \\ x(s) &= \int_0^s \sin(\chi(u))du \\ y(s) &= \int_0^s \cos(\chi(u))du \\ & 0 \le s \le \ell \end{aligned} \tag{4.206}$$

Clothoids The curvature of a clothoid curve is proportional to the length of the path as a curve element $\kappa(s) = \sigma_c s$, where σ_c is a real constant called the sharpness of the Clothoid. In the xy plane, this is a well known path in mobile robotics: clothoid curve [69, 101].

An elementary path is considered as a concatenation of two equal piecewise clothoids.

$$\begin{aligned} \kappa(s) &= \sigma_c s \quad \forall s \in \left[0, \frac{\ell}{2}\right] \\ \kappa(s) &= \sigma_c(\ell - s) \quad \forall s \in \left[\frac{\ell}{2}, \ell\right] \end{aligned} \tag{4.207}$$

Furthermore, two configurations in the plane can be obtained with two different elementary paths to form a bi-elementary path. In particular to link two successive way-points $P_i = (x_i, y_i, \chi_i)^T$, $P_f = (x_f, y_f, \chi_f)^T$, it is necessary to calculate the split configurations $P_s = (x_s, y_s, \chi_s)^T$ which is a symmetric configurations with respect to the start and end configurations. For symmetrical configurations, the following relations are verified

$$\begin{aligned} &\kappa\left(\frac{\ell}{2}\right) = -\kappa\left(-\frac{\ell}{2}\right) \\ &\chi\left(\frac{\ell}{2}\right) = \chi\left(-\frac{\ell}{2}\right) = \frac{\zeta}{2} \quad \Rightarrow \quad -\frac{1}{2} \le s \le \frac{1}{2} \\ &D(\zeta) = 2\int_0^{0.5} \sin(2\zeta u^2)du \end{aligned} \qquad \begin{aligned} \kappa(s) &= 4\zeta s \\ \chi(s) &= 2\zeta s^2 \\ x(s) &= \int_0^s \sin(2\zeta u^2)du \\ y(s) &= \int_0^s \cos(2\zeta u^2)du \end{aligned} \tag{4.208}$$

For non symmetrical configurations, one intermediate configuration symmetrical to both initial and final configurations must be found. It belongs to a circle. Then two portions of clothoids must be joined.

For a real clothoid, the following relations are used

1. homothetie $\ell = \frac{d}{D(\zeta)}$
2. rotation $\chi(s) \to \chi(s) + \varsigma$
3. translation $x(s) \to x(s) + x_0$ and $y(s) \to y(s) + y_0$

The loci of split configurations joining the start and end configurations with a bi-elementary path is a circle. Therefore, it exists an infinite set of solutions (bi-elementary paths with different lengths and curvatures) joining start and end configurations. In particular, the shortest bounded curvature path satisfying $\kappa \leq \kappa_{max}$ is of interest.

Spiral Cubic For symmetrical configurations

$$\begin{aligned}
D(\zeta) &= 2\int_0^{0.5} \cos\left(\zeta s\left(\frac{3}{2} - 2u^2\right)\right) du \\
\ell &= \frac{d}{D(\zeta)} \\
\kappa(s) &= \frac{6\zeta}{\ell^3}\left(\frac{\ell^2}{4} - s^2\right) \\
\chi(s) &= \frac{6\zeta}{\ell^3}\left(\frac{\ell^2}{4}s - \frac{s^3}{3}\right) \\
x(s) &= \int_0^s \sin(\chi(u))du \\
y(s) &= \int_0^s \cos(\chi(u))du
\end{aligned} \tag{4.209}$$

The locus of intermediate configurations for non symmetrical initial and final configurations are on

1. if P_i and P_f are parallel configurations then it is on the line

$$(x - x_i)(y - y_f) = (x - x_f)(y - y_i)$$

2. it not, the loci is a circle whose center is given by

$$P_c = \left(\frac{x_i + x_f + c(y_i + y_f)}{2}, \frac{y_i + y_f + c(x_i + x_f)}{2}\right)$$

where $c = (\tan \frac{\chi_i + \chi_f}{2})^{-1}$.

Polar Polynomials 2D Polar Polynomials are defined as

$$\begin{aligned}
x &= r(\chi)\sin\chi \\
y &= r(\chi)\cos\chi
\end{aligned} \tag{4.210}$$

The polar radius r is a polynomial function of the angle χ

$$r(\chi) = R\left(1 + \frac{\chi^2}{2} - \frac{\chi^3}{\chi_f - \chi_i} + \frac{\chi^4}{2(\chi_f - \chi_i)^2}\right)$$

Fig. 4.9 3D curve with linear torsion and curvature

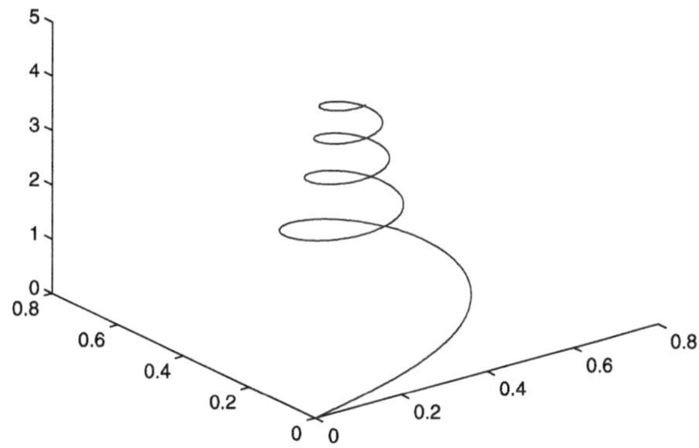

4.6.3.2 γ Angle Constant

With this assumption, the following relations are proposed:

$$
\begin{aligned}
dx &= \sin(\chi_0 + \chi_1 s + \chi_2 s^2)\cos(\gamma_0)ds \\
dy &= \cos(\chi_0 + \chi_1 s + \chi_2 s^2)\cos(\gamma_0)ds \\
dz &= \sin(\gamma_0)ds
\end{aligned}
\tag{4.211}
$$

Path in the x-y plane is shown in Fig. 4.9 (The units are in 1000 m). In 3D, this path has as curvature and torsion linear functions of s:

$$\kappa(s) = (\chi_1 + 2\chi_2 s)\cos(\gamma_0) \tag{4.212}$$

$$\tau(s) = (\chi_1 + 2\chi_2 s)\sin(\gamma_0) \tag{4.213}$$

Limitations κ_{max}, τ_{max} on these two parameters can be easily added.

$$(\chi_1 + 2L\chi_2) \leq \frac{\kappa_{max}}{\cos(\gamma_i)} \tag{4.214}$$

$$(\chi_1 + 2L\chi_2) \leq \frac{\tau_{max}}{\sin(\gamma_i)} \tag{4.215}$$

or

$$\chi_1 \leq \frac{\kappa_{max}}{\cos(\gamma_i)} \tag{4.216}$$

$$\chi_1 \leq \frac{\tau_{max}}{\sin(\gamma_i)} \tag{4.217}$$

This relationship must be respected

$$\chi_i + \chi_1 L + \chi_2 L^2 - \chi_f = 0 \tag{4.218}$$

Once the path is chosen, the temporal variation of the lighter than air robot motion has to be determined. On this path, an optimal control problem can be formulated using the following nominal force, with respect to the acceleration:

$$F^2 = u_1^2 + u_2^2 + u_3^2 \tag{4.219}$$

where

$$\begin{aligned} u_1 &= m\dot{V} + \frac{1}{2}C_D(M,\alpha)V^2 A_r \rho + mg\sin\gamma \\ u_2 &= mV\dot{\gamma} - \frac{1}{2}C_L(M,\alpha)V^2 A_r \rho \cos\sigma + mg\cos\gamma \\ u_3 &= mV\cos\gamma\dot{\chi} - \frac{1}{2}C_L(M,\alpha)V^2 A_r \rho \sin\gamma \end{aligned} \tag{4.220}$$

A time varying velocity can be obtained as a result of this optimization.

A generalization can be easily obtained for a quadratic variation of the curvature and torsion.

4.6.3.3 Angle γ as a Linear Function of s

In this case, the following relations are proposed:

$$\begin{aligned} dx &= \sin(\chi_0 + \chi_1 s)\cos(\gamma_0 + \gamma_1 s)ds \\ dy &= \cos(\chi_0 + \chi_1 s)\cos(\gamma_0 + \gamma_1 s)ds \\ dz &= \sin(\gamma_0 + \gamma_1 s)ds \end{aligned} \tag{4.221}$$

Using the initial and final configurations, the following relations must be respected:

$$\gamma_1 = \frac{z_f - z_i}{\cos\gamma_i - \cos\gamma_f} \tag{4.222}$$

and

$$\chi_1 = \frac{(z_f - z_i)(\chi_f - \chi_i)}{(\cos\gamma_i - \cos\gamma_f)(\gamma_f - \gamma_i)} \tag{4.223}$$

In 3D, this path has as curvature:

$$\kappa(s) = (\gamma_1^2 + \chi_1^2\cos^2(\gamma_0 + \gamma_1 s))^{-1/2} \tag{4.224}$$

The obtained flight path is depicted in Fig. 4.10 (The units are in 1000 m). A generalization in this direction would give a more complicated formulation of the curvature and torsion.

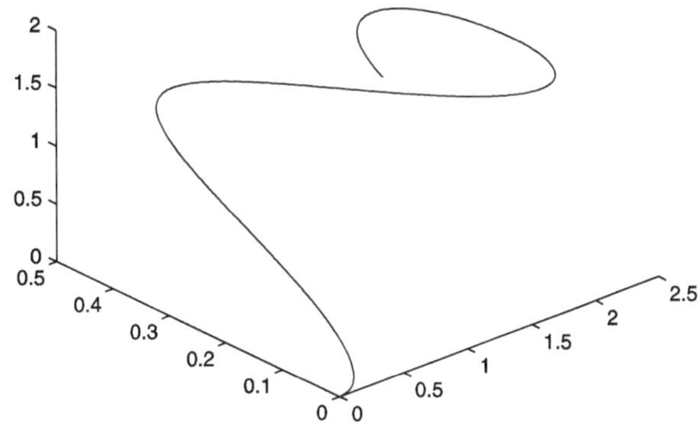

Fig. 4.10 3D curve with linear varying heading and flight path angles

4.6.4 Maneuvers Between Two Different Trims

The problem of maneuvers between two different trim trajectories is an important problem in motion planning for an aerial vehicle. Let's consider the case of joining a trim path to a non trim one, in the case of a constant flight angle γ_i. Let L_1 and L_2 be respectively the length of the first and the second part. The coordinates of the junction point (x_j, y_j, z_j, χ_j) have to be determined, by the resolution of a set of nonlinear equations.

$$x_j = x_i - \frac{\cos\gamma_i}{\chi_1}(\sin\chi_j - \sin\chi_0) \tag{4.225}$$

$$y_j = y_i + \frac{\cos\gamma_i}{\chi_1}(\cos\chi_j - \cos\chi_0) \tag{4.226}$$

$$z_j = z_i + L_1 \sin\gamma_i = z_f - L_2 \sin\gamma_i \tag{4.227}$$

The following simulation results are used to join a trim path (in blue) to a non trim path (in red) with constant torsion and linear curvature. The initial configuration for both paths is $x_0 = 0$ m; $y_0 = 0$ m; $z_0 = 100$ m; $\chi_0 = 0$ rad; $\gamma_0 = 0.1$ rad; For the trim path, the final configuration is $x_t = 900$ m; $y_t = 1400$ m; $z_t = 300$ m; $\chi_t = 2$ rad; $\gamma_t = 0.1$ rad. For the non trim path, the final configuration is $x_{nt} = 400$ m; $y_{nt} = 950$ m; $z_{nt} = 300$ m; $\chi_{nt} = 4$ rad; $\gamma_{nt} = 0.1$ rad.

Trim (shown by a thin line) and non trim (shown by a bold line) 3D paths are shown in Fig. 4.11 (The units are m). The trim trajectories have the advantage of facilitating the planning and control problems. A linear control technique with variable parameters could be sufficient to stabilize the vehicle in the neighborhood of trim conditions.

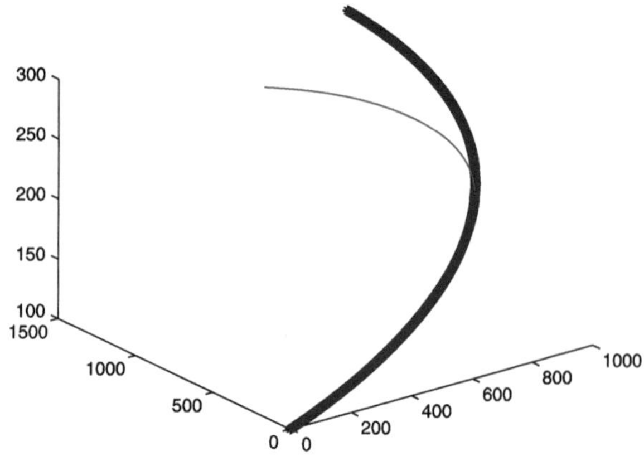

Fig. 4.11 (Color online) 3D trim (*thick*) and non trim (*fine*) paths with linear curvature

4.6.5 Frenet-Serret Approach

Let's propose the curvature and the torsion as continuous functions of the curvilinear abscissa s then find the corresponding path. Polynomial functions are a good example of continuous easy to deal with functions. So, let's take the curvature and torsion as [28]:

$$\kappa(s) = \sum_{i=0}^{n} a_i s^i \qquad \tau(s) = \sum_{j=0}^{m} b_j s^j \tag{4.228}$$

n and m represent respectively the order of curvature and torsion polynomials.

The differential equation describing the evolution of Φ can be written as:

$$\frac{d\Phi(s)}{ds} = \Phi(s) Sk \begin{pmatrix} \kappa \\ 0 \\ \tau \end{pmatrix} \tag{4.229}$$

Integrating this matrix differential equation gives:

$$\begin{gathered} \Phi(s) = \exp(\varpi(s)) \in SO(3) \\ \varpi(s) = \begin{pmatrix} \sum_{i=0}^{m} \frac{a_i}{i+1} s^{i+1} \\ 0 \\ \sum_{j=0}^{m} \frac{b_j}{j+1} s^{j+1} \end{pmatrix} \end{gathered} \tag{4.230}$$

Using Rodrigues formula to write a closed form solution

$$\Phi(s) = I_{3\times 3} + \frac{\sin\|\varpi(s)\|}{\|\varpi(s)\|}\varpi(s) + \frac{1-\cos\|\varpi(s)\|}{\|\varpi(s)\|}\varpi^2(s) \tag{4.231}$$

The tangent vector of the Frenet-Serret frame

$$T(s) = \Phi(s)\begin{pmatrix}1\\0\\0\end{pmatrix} = \begin{pmatrix} 1 - \varpi_3^2(s)\frac{1-\cos\sqrt{\varpi_1^2(s)+\varpi_3^2(s)}}{\varpi_1^2(s)+\varpi_3^2(s)} \\ \varpi_3(s)\frac{\sin\sqrt{\varpi_1^2(s)+\varpi_3^2(s)}}{\sqrt{\varpi_1^2(s)+\varpi_3^2(s)}} \\ \varpi_1(s)\varpi_3(s)\frac{1-\cos\sqrt{\varpi_1^2(s)+\varpi_3^2(s)}}{\varpi_1^2(s)+\varpi_3^2(s)} \end{pmatrix}$$

while the normal vector is

$$N(s) = \Phi(s)\begin{pmatrix}0\\1\\0\end{pmatrix} = \begin{pmatrix} -\varpi_3(s)\frac{\sin\sqrt{\varpi_1^2(s)+\varpi_3^2(s)}}{\sqrt{\varpi_1^2(s)+\varpi_3^2(s)}} \\ 1 - \left(\varpi_1^2(s) + \varpi_3^2(s)\right)\frac{1-\cos\sqrt{\varpi_1^2(s)+\varpi_3^2(s)}}{\varpi_1^2(s)+\varpi_3^2(s)} \\ \varpi_1(s)\frac{\sin\sqrt{\varpi_1^2(s)+\varpi_3^2(s)}}{\sqrt{\varpi_1^2(s)+\varpi_3^2(s)}} \end{pmatrix}$$

and the binormal vector is:

$$B(s) = \Phi(s)\begin{pmatrix}0\\0\\1\end{pmatrix} = \begin{pmatrix} \varpi_1(s)\varpi_3(s)\frac{1-\cos\sqrt{\varpi_1^2(s)+\varpi_3^2(s)}}{\varpi_1^2(s)+\varpi_3^2(s)} \\ -\varpi_1(s)\frac{\sin\sqrt{\varpi_1^2(s)+\varpi_3^2(s)}}{\sqrt{\varpi_1^2(s)+\varpi_3^2(s)}} \\ 1 - \varpi_1^2(s)\frac{1-\cos\sqrt{\varpi_1^2(s)+\varpi_3^2(s)}}{\varpi_1^2(s)+\varpi_3^2(s)} \end{pmatrix}$$

Finally, the 3D curve is obtained via the Fresnel integrals

$$C(s) = \int T(u)du \tag{4.232}$$

Doing a parallel with planar curves, three cases can be studied:

- constant curvature and torsion
- curvature and/or torsion linear function of s
- curvature and/or torsion quadratic function of s

4.6.6 Pythagorean Hodograph

The above presented curves are defined in terms of the Fresnel integrals which are not rational and not exactly expressible as a non uniform rational B-splines

(NURBS). As an alternative, the cubic Bezier and Pythagorean curves can be explored. The Pythagorean Hodograph curves are introduced to provide exact solutions to a number of basic computational problems that arise in robot path planning. These include analytic reduction of the arc length and bending energy integrals, construction of rational offset (parallel) curves, formulation of real time interpolator for motion control applications; determination of rotation minimizing frames for specifying orientations of rigid bodies along spatial paths.

A polynomial space curve is said to Double Pythagorean Hodograph (DPH) if the Frenet frame, the curvature and the torsion have all a rational dependence on the curve parameter [68, 190].

4.6.6.1 2D Pythagorean Hodograph

In 2D, a Pythagorean Hodograph $r(t) = (x(t), y(t))$ is a polynomial curve whose tangents $\dot{x}(t)$ and $\dot{y}(t)$ satisfies

$$\dot{x}^2(t) + \dot{y}^2(t) = \sigma^2(t) \tag{4.233}$$

for some polynomial $\sigma(t)$ where $\dot{x}(t) = \frac{dx}{dt}$ and $\dot{y}(t) = \frac{dy}{dt}$. From the principles of differential geometry, the path length s and parametric speed $\dot{s}$ of a parametric curve are given by:

$$s = \int_{t_1}^{t_2} \|\dot{r}(t)\| \, dt = \int_{t_1}^{t_2} \sqrt{\dot{x}^2(t) + \dot{y}^2(t)} dt \tag{4.234}$$

If the sum of square of the tangents $\dot{x}(t)$, $\dot{y}(t)$ could be represented by perfect square of a single polynomial, it leads to two advantages:

- The radical form for calculating the path length is eliminated
- The parametric speed of the curve is simply a polynomial function of the parameter t.

This is achieved by forming the hodograph of the curve $r(t)$ of the form:

$$\begin{aligned} \dot{x}(t) &= w(t)\left(u^2(t) - v^2(t)\right) \\ \dot{y}(t) &= 2w(t)u(t)v(t) \end{aligned} \tag{4.235}$$

Equation (4.233) becomes

$$\sqrt{\dot{x}^2(t) + \dot{y}^2(t)} = \sigma(t) = w(t)\left(u^2(t) + v^2(t)\right) \tag{4.236}$$

where $u(t)$, $v(t)$, $w(t)$ and $\sigma(t)$ are non zero real polynomials, satisfying $gcd(u(t), v(t)) = 1$ (greater common divisor). Now, Eq. (4.234) becomes

$$s = \int_{t_1}^{t_2} \sigma(t) dt \tag{4.237}$$

Depending on the order of the polynomials $u(t)$, $v(t)$, $w(t)$ the Pythagorean Hodograph curve can be either cubic, quartic or quintic. The expressions for unit tangent T, unit normal N and curvature κ of a Pythagorean Hodograph curve are:

$$T = \frac{(u^2 - v^2, 2uv)}{u^2 + v^2} \qquad N = \frac{(2uv, v^2 - u^2)}{u^2 + v^2}$$
$$\kappa = \frac{(2\,(u\dot{v} - v\dot{u})\,, 2uv)}{w(u^2 + v^2)} \tag{4.238}$$

The off-set curve $r_0(t)$ of the curve $r(t)$ at a distance $\pm d$ is

$$r_0(t) = r(t) \pm dN(t)$$

4.6.6.2 3D Pythagorean Hodograph

In 3D, the Pythagorean Hodograph is defined by a velocity

$$\dot{s} = \sqrt{\dot{x}^2(t) + \dot{y}^2(t) + \dot{z}^2(t)}$$

polynomial in t. The arc length of the Pythagorean Hodograph curve can be computed precisely by evaluating a polynomial. For the hodograph $\dot{r} = (\dot{x}(t), \dot{y}(t), \dot{z}(t))$, it is necessary and sufficient that its components be expressible in terms of polynomials $u(t), v(t), p(t), q(t)$ in the form

$$\begin{aligned}
\dot{x}(t) &= u^2(t) + v^2(t) - p^2(t) - q^2(t)\\
\dot{y}(t) &= 2\,(u(t)q(t) + v(t)p(t))\\
\dot{z}(t) &= 2\,(v(t)q(t) - u(t)p(t))\\
\sigma(t) &= u^2(t) + v^2(t) + p^2(t) + q^2(t)
\end{aligned} \tag{4.239}$$

If $u(t), v(t), p(t), q(t)$ are all constants, the hodograph is a single point specifying a uniformly parameterized straight line.

The simplest non trivial Pythagorean Hodograph are cubic, they correspond to segments of non circular helices with a constant ration $\frac{\kappa}{\tau}$. They may be characterized by certain geometrical constraints on these Bezier control polygons.

$$r(t) = \sum_{k=1}^{3} b_k \binom{3}{k} (1-t)^{3-k}\, t^k \quad t \in [0, 1] \tag{4.240}$$

To guarantee sufficient shape flexibility, fifth order Pythagorean Hodograph curve may be employed. The construction of spatial Pythagorean Hodograph fifth order is described, in Bernstein-Bezier form

$$r(t) = \sum_{k=1}^{5} b_k \binom{5}{k} (1-t)^{5-k}\, t^k \quad t \in [0, 1] \tag{4.241}$$

where $b_k = (x_k, y_k, z_k)$ are control points, whose vertices define the control polygon or Bezier polygon, t is a parameter and $k = 0 \dots 5$. For PH fifth orders, $|r' \times r''| = \sigma^2 \iota$ and $\iota = (up' - u'p + vq' - v'q)^2 + (uq' - u'q + v'p - vp')^2$.

For PH fifth order, $(r' \times r'').r'''$ is of degree 6 while σ, ι are both in degree 4 in t. For a PH curve with $\iota = \omega^2$, the Frenet frame vectors and the curvature and torsion functions are given by the rational expressions

$$\begin{aligned} T = \frac{r'}{\sigma} \qquad N = \frac{\sigma r'' - \sigma' r'}{\sigma \omega} \qquad B = \frac{r' \times r''}{\sigma \omega} \\ \kappa = \frac{\omega}{\sigma^2} \qquad \tau = \frac{(r' \times r'').r'''}{\sigma^2 \omega^2} \end{aligned} \tag{4.242}$$

Hence, the PH curves may be regarded as the complete set of polynomial curves that have rational Frenet frame.

Let the initial and final configurations be $(x_i, y_i, z_i, \chi_i, \gamma_i)^T$ and $(x_f, y_f, z_f, \chi_f, \gamma_f)^T$. The four control points of the Bezier polygon are calculated by first order Hermite interpolation as follows:

$$\begin{aligned} b_0 &= (x_i, y_i, z_i) \\ b_5 &= \left(x_f, y_f, z_f\right) \\ b_1 &= b_0 + \frac{1}{5} (\sin \chi_i \cos \gamma_i, \cos \chi_i \cos \gamma_i, \sin \gamma_i) \\ b_4 &= b_5 - \frac{1}{5} \left(\sin \chi_f \cos \gamma_f, \cos \chi_f \cos \gamma_f, \sin \gamma_f\right) \end{aligned} \tag{4.243}$$

The control points b_0, b_1, b_4, b_5 are fixed. Now the problem is reduced to finding the control points b_2, b_3. Both polynomials curves are given by

$$\begin{aligned} u(t) &= u_0(1-t)^2 + 2u_1(1-t) + u_2 t^2 \\ v(t) &= v_0(1-t)^2 + 2v_1(1-t) + v_2 t^2 \\ p(t) &= p_0(1-t)^2 + 2p_1(1-t) + p_2 t^2 \\ q(t) &= q_0(1-t)^2 + 2q_1(1-t) + q_2 t^2 \end{aligned} \tag{4.244}$$

Knowing that

$$\int \binom{n}{k} (1-t)^{n-k} t^k dt = \frac{1}{n+1} \sum_{i=k+1}^{n+1} (1-t)^{n+1-i} t^i \tag{4.245}$$

The set of equations to be solved for the control points b_2, b_3 results in four solutions. Among these four paths, only one has an acceptable shape that is without twists and cusps. This path will be used as reference and is identified by calculating the bending energy of the curve and choosing the path whose energy is minimal.

4.6.7 η^3 Splines

The η^3 Splines can represent any seventh order polynomial curve with third order geometric continuity denoted as G^3 continuity: continuous in position, curvature and derivative of the curvature [170].

The 2D problem can be formulated as: Determine the minimal order polynomial curve which interpolates two given configurations $P_i = (x_i, y_i, \chi_i, \kappa_i, \kappa_i')$ and $P_f = (x_f, y_f, \chi_f, \kappa_f, \kappa_f')$. The 7th order is the minimal order polynomial curve interpolating such configurations.

The solution proposed for the above interpolating problem is given by a 7th order polynomial curve $P(u) = (x(u), y(u)), u \in [0, 1]$ defined below:

$$\begin{aligned}
x(u) &= \alpha_0 + \alpha_1 u + \alpha_2 u^2 + \alpha_3 u^3 + \alpha_4 u^4 + \alpha_5 u^5 + \alpha_6 u^6 + \alpha_7 u^7 \\
y(u) &= \beta_0 + \beta_1 u + \beta_2 u^2 + \beta_3 u^3 + \beta_4 u^4 + \beta_5 u^5 + \beta_6 u^6 + \beta_7 u^7
\end{aligned} \tag{4.246}$$

The polynomial coefficients are detailed by the following closed-form expressions, by solving a nonlinear equation system associated to the end point interpolation conditions:

$$\begin{aligned}
\alpha_0 &= x_i \\
\alpha_1 &= \eta_1 \cos\chi_i \\
\alpha_2 &= \frac{1}{2}\eta_3 \cos\chi_i - \frac{1}{2}\eta_1^2\kappa_i \sin\chi_i \\
\alpha_3 &= \frac{1}{6}\eta_5 \cos\chi_i - \frac{1}{6}\left(\eta_1^3\kappa_i' + 3\eta_1\eta_3\kappa_i\right)\sin\chi_i \\
\alpha_4 &= 35(x_f - x_i) - \left(20\eta_1 + 5\eta_3 + \frac{2}{3}\eta_5\right)\cos\chi_i \\
&\quad + \left(5\eta_1^2\kappa_i + \frac{2}{3}\eta_1^3\kappa_i' + 2\eta_1\eta_3\kappa_i\right)\sin\chi_i \\
&\quad - \left(15\eta_2 - \frac{5}{2}\eta_4 + \frac{1}{6}\eta_6\right)\cos\chi_f \\
&\quad - \left(\frac{5}{2}\eta_2^2\kappa_f - \frac{1}{6}\eta_2^3\kappa_f' - \frac{1}{2}\eta_2\eta_4\kappa_f\right)\sin\chi_f \\
\alpha_5 &= -84(x_f - x_i) + (45\eta_1 + 10\eta_3 + \eta_5)\cos\chi_i \\
&\quad - \left(10\eta_1^2\kappa_i + \eta_1^3\kappa_i' + 3\eta_1\eta_3\kappa_i\right)\sin\chi_i \\
&\quad + \left(39\eta_2 - 7\eta_4 + \frac{1}{2}\eta_6\right)\cos\chi_f \\
&\quad + \left(7\eta_2^2\kappa_f - \frac{1}{2}\eta_2^3\kappa_f' - \frac{3}{2}\eta_2\eta_4\kappa_f\right)\sin\chi_f
\end{aligned} \tag{4.247}$$

$$\alpha_6 = 70(x_f - x_i) - \left(36\eta_1 + \frac{15}{2}\eta_3 + \frac{2}{3}\eta_5\right)\cos\chi_i$$
$$+ \left(\frac{15}{2}\eta_1^2\kappa_i + \frac{2}{3}\eta_1^3\kappa_i' + 2\eta_1\eta_3\kappa_i\right)\sin\chi_i$$
$$- \left(34\eta_2 - \frac{13}{2}\eta_4 + \frac{1}{2}\eta_6\right)\cos\chi_f$$
$$- \left(\frac{13}{2}\eta_2^2\kappa_f - \frac{1}{2}\eta_2^3\kappa_f' - \frac{3}{2}\eta_2\eta_4\kappa_f\right)\sin\chi_f$$

$$\alpha_7 = -20(x_f - x_i) + \left(10\eta_1 + 2\eta_3 + \frac{1}{6}\eta_5\right)\cos\chi_i$$
$$- \left(2\eta_1^2\kappa_i + \frac{1}{6}\eta_1^3\kappa_i' + \frac{1}{2}\eta_1\eta_3\kappa_i\right)\sin\chi_i$$
$$+ \left(10\eta_2 - 2\eta_4 + \frac{1}{6}\eta_6\right)\cos\chi_f$$
$$+ \left(2\eta_2^2\kappa_f - \frac{1}{6}\eta_2^3\kappa_f' - \frac{1}{2}\eta_2\eta_4\kappa_f\right)\sin\chi_f$$

More informations can be found in [170]. The real vector $\eta_v = (\eta_1, \eta_2, \eta_3, \eta_4, \eta_1, \eta_5, \eta_6)^T$ can be freely selected and influence the path shape without violating the end-point interpolating conditions. This solution represents a family of curves that depend on a symmetric parameterization induced by the chosen η_v vector. Specifically, parameters η_1, η_3, η_5 influence the curve at its beginning while the parameters η_2, η_4, η_6 affect the curve ending. Parameters η_1, η_2 can be interpreted as velocity parameters while parameters $\eta_3, \eta_4, \eta_5, \eta_6$ are twist parameters depending on curve accelerations and jerks at the end-points.

Acting on the shaping parameters vector η, a wide variety of curves satisfying the boundary conditions can be obtained. This suggests choosing η to generate optimal curves. Different optimality criteria may be chosen depending on the mission of the lighter than air robot.

The role of the trajectory generator is to generate a feasible time trajectory for the aerial robot. Once the path has been calculated in the Earth fixed frame, motion must be investigated and reference trajectories determined taking into account actuators constraints.

4.7 Conclusions

The fundamentals of flight are in general: straight and level flight (maintenance of selected altitude), ascents and descents, level turns and wind drift correction. In the first section of this chapter, trim trajectories are presented. Trim is concerned with the ability to maintain flight equilibrium with controls fixed. A trimmed flight condition is defined as one in which the rate of change (of magnitude) of the lighter than air state vector is zero (in the body-fixed frame) and the resultant of the applied forces and moments is zero.

The second part of this chapter addresses the problem of characterizing continuous paths in 3D, taking into account the under-actuation. Three differential algebraic equations must be solved as there is six degrees of freedom and three inputs. The constraint on the yaw angle is in the form of a generalized logistic equation while the other constraints are differential algebraic equations in v and w, when the variations of the longitudinal velocity u, and the pitch and roll angles are imposed.

In the third section, an algorithm for 2D and 3D open-loop path planning is derived for the system presented in the previous section. The idea is to use the structure and to apply simple bang-bang controls in the planning. The amount of control available is a concern in the planning for this system due to the drift term. The class of bang-bang controls is often a sufficiently rich class of controls for analysis of nonlinear systems. This simple class of controls makes it possible to integrate the equations forward in a simple manner. Zermelo navigation problem is considered in the sequel.

Then the problem of characterizing continuous paths in 3D is addressed with constraints on curvature and torsion, as consequence from limitations on thrust and velocity. Parametric paths are investigated, depending on the initial and final configurations. Smoother paths can be obtained by asking for the continuity of the derivatives of the path curvature and torsion. This section addresses the problem of characterizing continuous paths in 3D. Parametric paths with given curvature and torsion are investigated. Two particular cases are studied: constant, linear and quadratic variation of the heading angle versus the curvilinear abscissa, with the assumption of a constant or linear variation of the flight path angle, to consider different kinds of maneuvers. Maneuvers should be kept only to join two trim flight paths. Finally, some parametric curves such as polynomials, Pytaghorean Hodograph, ^{3} splines are presented.

Trajectory planning incorporates dynamics into planning processes. Depending on the mission, time variable velocity must be considered, giving more flexibility to the trajectory generator, with respect to the limitations on actuators, on curvature and torsion. This is the topic of future work for non trim trajectories.

Chapter 5
Control

Abstract The control methods implemented on lighter than air robots lie in two categories: traditional control methods and advanced control methods. The traditional control methods achieve autonomous control goals via classical control algorithms. These control methods have the advantage of being easily implemented and providing reliable control performance while the weaknesses include the costs of computation to model the system and tuning the control parameters. The most basic nonlinear control laws are the On-off control and Gain scheduling. Most of the advanced control methods are faced with highly nonlinear and time varying control system, in which it is difficult to obtain an accurate dynamic model of the LTAR and the environment. Several control methods have been developed such as back stepping control, robust control, model-prediction control and other intelligent control methods.

5.1 Introduction

Linear control design techniques have been used for flight control problems for many years [48, 108]. One of the reasons why airships can be controlled quite well by linear controllers is that they behave almost linearly through most of their flight envelope. However, when the airship is required to pass through a highly nonlinear dynamic region or when other complicated control objectives are set, it is difficult to obtain practical controllers based on linear design techniques. The sources of nonlinearities are the aerodynamic forces generated at low airspeeds and high angles of attack (especially when wind disturbances are present); trajectory constraints imposed due to proximity of ground; kinematic nonlinearities when active maneuvering is required; actuator saturations...

Advanced control methods are becoming more popular for lighter than air robot as these control methods are mainly developed to improve the control performance of the lighter than air robot in a complex and unstable flight environment [1, 9, 33, 37–39, 106, 111, 134, 151]. In contrast to the linear systems, the characteristics of the nonlinear systems are not simply classified and there are no general methods comparable in power to those of linear analysis. Nonlinear techniques are quite often designed for individual cases [4, 50, 88, 98, 99, 197, 216, 230].

Y. Bestaoui Sebbane, *Lighter than Air Robots*,
Intelligent Systems, Control and Automation: Science and Engineering 58,
DOI 10.1007/978-94-007-2663-5_5, © Springer Science+Business Media B.V. 2012

A completely different approach is to enable applicability of the well known linear control methods to control nonlinear systems. This can be achieved using nonlinear dynamic inversion [150, 209]. This process, also known as feedback linearization, involves on line approximate linearization of a nonlinear plant via feedback.

Another control technique is variable structure (also known as sliding mode) control [31, 75–77]. In this approach, a hyper surface (in state space) called sliding surface or switching surface is selected so that the system trajectory exhibits desirable behavior when confined to this hyper surface. Depending on whether the current state is above or below the sliding surface, a different control gain is applied. Unlike gain scheduling, the method involves high speed switching to keep the system on the sliding surface. Adaptive control term covers a set of various control techniques that are capable of on line adaptation. The applications of adaptive control is generally biased toward control for large time scales so that the controller has sufficient time to learn how to behave. This makes the relatively short-time recovery process unsuitable for on line adaptation.

Based on the airship dynamic model research, Azinheira [9] has done a specific research on hovering control of the airship. He uses the visual servo control and back stepping control technology to realize the hovering tasks in the outdoor environment. Takaya [205] developed a PID controller to control the landing motion for an indoor blimp robot. Moutinho in [150] used the dynamic inversion to realize robust control method in their autonomous airship of the Aurora project. A dynamic inversion controller was implemented with desired dynamics given by a linear optimal compensator. The stability analysis of the nonlinear system is done applying Lyapunov's stability theory. Fukao in [75] implemented inverse optimal tracking control to improve the robustness of an autonomous airship control system. Back stepping control and model prediction control are efficient and well discussed control algorithms. Hygounenc [89, 90] deals with very low perturbations, for each phase, a reduced model is determined and a controller is designed on the basis of back stepping procedures.

The Fault Detection and Isolation (FDI) technique is one of the important processes in a redundancy management system. Fault detection refers to the decision of whether a fault occurred or not, and fault isolation is the process of finding and excluding the failed sensor. Currently, the FDI techniques have been developed using hardware redundancy or analytical redundancy. Hardware redundancy usually makes the system complicated because multiple redundant sensors are used for cross channel monitoring. On the other hand, analytical redundancy uses the information from the system's mathematical model [44, 49, 56, 66, 70, 75, 146].

5.2 Linear Control

The complexity of the lighter than air robot non linear dynamic equations justifies the search for a linear version, also important in order to analyze and evaluate the characteristics of the vehicle dynamics. A linear dynamics model is formulated to allow a quantitative assessment of the flight stability and the response to control inputs of the lighter than air robot. To derive the linear dynamics model, the first step is to introduce a reference equilibrium state. The second step is to write the equations for the small disturbance from trim. In small perturbations about the trimmed equilibrium, it is also possible to decouple longitudinal from lateral motion. Once the state matrix is obtained, the lighter than air robot flying and handling qualities is characterized by the eigenvalues and eigenvectors. For a stable lighter than air robot the real parts of the eigenvalues must be negative. Each element of the eigenvectors denotes the magnitude and phase of the response of a particular state variable relative to other states. The purpose of this section is to show some recent approaches in the context of aerial robots research in order to highlight the potential advantages of control methods. It is an efficient way to analyze and improve the control in the aerial robots research. Many different control tasks could be considered for a lighter than air robot to achieve a successful flight test. Motion control is the most developed and discussed issue in practical applications. It is used to autonomously and efficiently control certain flight motions, such as hovering, taking off, landing motions. . . A common approach to deal with nonlinearities is linearization about a reference trajectory.

There are two good reasons for algebraically deriving the small-perturbation equations.

- the aerodynamic parameters needed for the linear equations can be estimated relatively quickly
- the algebraic small-perturbation equations provide a great deal of insight into the relative importance of the various aerodynamic derivatives under different flight conditions and their effect on the stability of the lighter than air robot motion.

There are some constraints on the evolution of the state of the system to ensure safe operations such as the flight envelope constraints closely related to the lighter than air robot and its dynamics or constraints on the state and the input such as fuel consumption.

5.2.1 Linear Formulation in Cruising Flight

The motion control tasks are important in analyzing dynamic models of the lighter than air robot. Linearization is the procedure in which a set of nonlinear differential equations is approximated by a linear set. There are two main reasons for the importance of linearization.

- there are many good design and analysis tools for linear systems
- if the small signal linear model of a system is stable, then there will exist a region, which may be small, within which the nonlinear system will be stable.

Thus, in general, the first technique of nonlinear system analysis and design is to obtain a linear approximation and design a controller for it. Simulation can then be used to explore the quality of this design when used with the nonlinear system model.

When it is assumed that motion of the lighter than air robot is constrained to small perturbations about the trimmed equilibrium flight condition then the model may be considerably simplified. In particular, the products and squares of small perturbation variables u, v, w, p, q, r become negligibly small. In small perturbations, it is also reasonable to assume decoupled longitudinal and lateral motion and consequently all coupling derivatives may be omitted from the equations.

5.2.1.1 Linear Approximation

Let's choose a state vector defined as $X = (u, w, q, \theta, v, p, r, \phi)^T$. The first four states describe the longitudinal motion and the last four the lateral motion. U is the control input vector, depending on the flight mode: hover or cruising. At low airspeeds, the propellers vectoring angle is necessary in order to compensate for the loss of lift force from aerodynamics. For very low airspeeds, the model is essentially that of aerostatic forces, with the weight excess being compensated by the propellers vectoring. For high airspeeds, the propellers vectoring is not necessary and the model is essentially that of aerodynamic forces. In the transition from low to high vectoring angle, there exists a compensation from both aerodynamic forces and propellers vectoring. For these reasons, this should be the most difficult region to be corroborated in the model validation process and in the control design [150].

To derive the linear dynamics model, the first step is to introduce a reference equilibrium or a trim state $\bar{X}$, along which the system will be linearized and the corresponding control $\bar{U}$.

If the lighter than air robot is in equilibrium flight then

$$\bar{X}_e = (u_e, w_e, q_e, \theta_e, v_e, p_e, r_e, \phi_e)^T.$$

The second step is to write the equations for the small disturbance from equilibrium as

$$\dot{X} = \mathbf{f}(\bar{X}_e + \Delta X, \bar{U}_e + \Delta U) \approx A\Delta X + B\Delta U \tag{5.1}$$

where $A = \frac{\partial \mathbf{f}}{\partial X}$ is the state matrix of the Jacobian of function f with respect to X with a similar definition of the control input matrix $B = \frac{\partial \mathbf{f}}{\partial U}$. These Jacobians are evaluated for the trim trajectories and control.

The matrices A and B can be evaluated numerically at $\bar{X}_e, \bar{U}_e$ by finite differences of the nonlinear differential equation. Once the matrix A is obtained, the stability of the lighter than air robot can be characterized by the eigenvalues and eigenvectors of A. They can be either distinct and real, representing non oscillatory modes, or complex conjugate as $\lambda_{1,2} = \sigma \pm j\omega_0$ representing oscillatory modes. For an oscillatory mode, the natural frequency and damping ratio can be evaluated from σ, ω_0. For a stable lighter than air robot, the real parts of all the eigenvalues must be negative. The eigenvectors represent the relationship of the elements of the state variables in the corresponding mode. Each element of an eigenvector denotes the magnitude and phase of the response of a particular state variable relative to other states [48, 201]. Once the matrices A and B are obtained, the control responses in the frequency domain can be computed.

The state matrix A can be partitioned in four distinct sub-matrices as:

$$A = \begin{pmatrix} A_{long} & 0 \\ 0 & A_{lat} \end{pmatrix} \tag{5.2}$$

where some of the elements of the non diagonal blocks are not exactly zero but by of much smaller magnitude than the elements in the 4×4 matrices A_{long}, A_{lat}. Therefore, the longitudinal and lateral motions can be decoupled.

5.2.1.2 Longitudinal Equations

For a cruising flight, the linearized longitudinal decoupled equations of motion may therefore be written as [108]:

$$\begin{aligned} &m_x\dot{u} + (ma_z - \dot{X}_{\dot{q}})\dot{q} \\ &\quad = X_e + \dot{X}_u u + \dot{X}_w w + (\dot{X}_q - m_z W_e) q + \dot{X}_\delta (\delta_e + \delta_r) + \dot{X}_t \delta_t \\ &\qquad + T_e - (mg - B)(\sin\theta_e + \theta\cos\theta_e) \end{aligned} \tag{5.3}$$

$$\begin{aligned} &m_z\dot{w} - (ma_x + \dot{Z}_{\dot{q}})\dot{q} \\ &\quad = Z_e + \dot{Z}_u u + \dot{Z}_w w + (\dot{Z}_q - m_x u_e) q + \dot{Z}_{\delta_e}\delta_e \\ &\qquad - (mg - B)(\cos\theta_e - \dot{\theta}\sin\theta_e) \end{aligned} \tag{5.4}$$

$$\begin{aligned} &J_y\dot{q} + (ma_x - \dot{M}_u)\dot{u} - (ma_x + \dot{M}_w)\dot{w} \\ &\quad = -\theta\big((mga_z + Bb_z)\cos\theta_e - (mga_x + Bb_x)\sin\theta_e\big) \\ &\qquad - (mga_z + Bb_z)\sin\theta_e - (mga_x + Bb_x)\cos\theta_e \end{aligned} \tag{5.5}$$

Longitudinal Trim Equations The condition for longitudinal trim may be deduced by noting that in trimmed equilibrium the perturbation variables are all zero and Eqs. (5.3, 5.4, 5.5) reduce to

$$\begin{aligned} X_e + T_e - (mg - B)\sin\theta_e &= 0 \\ Z_e + (mg - B)\cos\theta_e &= 0 \\ M_e + T_e d - (mga_x + Bb_x)\cos\theta_e - (mga_z + Bb_z)\sin\theta_e &= 0 \end{aligned} \tag{5.6}$$

Trim is thus achieved by adjusting thrust T_e, lift force B and center of volume b_x, b_z simultaneously.

Small Perturbations in Longitudinal Equations The linearized longitudinal equations of motion describing small perturbations about the trim state follow when the trim terms, which sum to zero are removed from Eqs. (5.3, 5.4, 5.5). Writing the resulting equations in state space form,

$$\dot{X}_{long} = A_{long} X_{long} + B_{long} U_{long} \tag{5.7}$$

where

$$\begin{aligned} X_{long}^T &= \begin{pmatrix} u & w & q & \theta \end{pmatrix} \\ U_{long}^T &= \begin{pmatrix} \delta_e & \delta_r \end{pmatrix} \end{aligned} \tag{5.8}$$

The state space matrix A_{long} and the control matrix B_{long} are thus given by the following relations

$$A_{long} = \begin{pmatrix} \dot{x}_u & \dot{x}_w & (\dot{x}_q - m_z W_e) & -(mg - B)\cos\theta_e \\ \dot{z}_u & \dot{z}_w & (\dot{z}_q + m_x U_e) & -(mg - B)\sin\theta_e \\ \dot{M}_u & \dot{M}_w & (\dot{M}_q - ma_x U_e - ma_z W_e) & -(mga_z + Bb_z)\cos\theta_e + (mga_x \sin\theta_e) \\ 0 & 0 & 1 & 0 \end{pmatrix}$$

$$B_{long} = \begin{pmatrix} x_\delta & x_t \\ z_\delta & 0 \\ m_\delta & m_t \\ 0 & 0 \end{pmatrix} \tag{5.9}$$

Classical controllability studies can be made with this linear formulation. If the Kalman condition $\text{rank}(B_{long}, A_{long} B_{long}, \dots, A_{long}^3 B_{long}) = 4$ is verified, the system is controllable.

5.2.1.3 Lateral Equations

The linearized lateral equations of motion may be developed similarly:

$$\begin{aligned} & m_y \dot{v} - \left(ma_z + \dot{Y}_{\dot{p}}\right)\dot{p} + \left(ma_x - \dot{Y}_{\dot{r}}\right)\dot{r} \\ &\quad = Y_e + \dot{Y}_v v + (\dot{Y}_p + m_z W_e)p + (\dot{Y}_r - m_x U_e)r + \dot{Y}_{\delta_r}\delta_r \\ &\quad\quad + (mg - B)\phi\cos\theta_e \end{aligned} \tag{5.10}$$

$$\begin{aligned} & J_z \dot{r} - J_{xz}\dot{p} + \left(ma_x - \dot{N}_{\dot{v}}\right)\dot{v} \\ &\quad = N_e + \dot{N}_v v + (\dot{N}_p + ma_x W_e)p + (\dot{N}_r - ma_x U_e)r + \dot{N}_{\delta_r}\delta_r \\ &\quad\quad + (mga_x + Bb_x)\phi\cos\theta_e \end{aligned} \tag{5.11}$$

$$\begin{aligned} & J_x \dot{p} - J_{xz}\dot{r} - \left(ma_z - \dot{L}_{\dot{v}}\right)\dot{v} \\ &\quad = L_e + \dot{L}_v v + (\dot{L}_p - ma_x W_e)p + (\dot{L}_r + ma_z U_e)r \\ &\quad\quad - (mga_z + Bb_z)\phi\cos\theta_e \end{aligned} \tag{5.12}$$

Trim Lateral Equations The condition for lateral trim may be deduced by noting that in trimmed equilibrium the perturbation variables are all zero and Eqs. (5.10,

5.11, 5.12) reduce to:

$$Y_e = L_e = N_e \tag{5.13}$$

The residual lateral force, yaw and roll moments are all zero.

Small Perturbations in Lateral Equations The linearized lateral equations of motion describing small perturbations about the trim state follow when the trim terms, which sum to zero are removed from Eqs. (5.10), (5.11), (5.12). Writing the resulting equations in state space form,

$$\dot{X}_{lat} = A_{lat} X_{lat} + B_{lat} U_{lat} \tag{5.14}$$

where

$$\begin{aligned} X_{lat}^T &= \begin{pmatrix} v & p & r & \phi \end{pmatrix} \\ U_{lat}^T &= \delta_r \end{aligned} \tag{5.15}$$

The state space matrix A_{lat} and the control matrix B_{lat} are thus given by the following relations

$$A_{lat} = \begin{pmatrix} y_v & y_p & y_r & y_\phi \\ l_v & l_p & l_r & z_\phi \\ n_v & n_p & n_r & n_\phi \\ 0 & 0 & 1 & 0 \end{pmatrix} \qquad B_{lat} = \begin{pmatrix} u_\delta \\ 0 \\ n_\delta \\ 0 \end{pmatrix} \tag{5.16}$$

If the Kalman condition $\text{rank}(B_{lat}, A_{lat}B_{lat}, \ldots, A_{lat}^3 B_{lat}) = 4$ is verified, the system is controllable.

5.2.1.4 Lateral Under-actuation

If the objective is hovering above the ground, the wind disturbance appears both as:

- a positive factor, which will help to control the lighter than air robot due to the increased authority of the tail control surfaces
- a drawback, producing a mostly horizontal force that needs to be balanced by a lighter than air robot actuator. This is only possible using the longitudinal forces and aligning the vehicle to reduce the drag forces.

As a consequence, and in order to avoid the saturation occurrences resulting from the reduced lateral controllability, along with the definition of suitable saturation limits, it is necessary to correct the control objective. In the presence of wind, the wind heading provides a non arbitrary yaw reference. This way, the airship may align itself face to the wind, and the lateral force input may vanish in stationary conditions.

Remark 5.1 The wind heading must be available. This value may be estimated using the measurement of the airship velocity as well as airspeed and side slip angle β.

5.2.2 Flying and Handling Qualities

A linear dynamics model is now formulated to allow a quantitative assessment of the flight stability and the frequency responses to control inputs of the lighter than air robot. Analytical linear dynamics models have been used to study airship stability [48, 64, 108, 113, 131, 182, 196, 212], the aerodynamics loads in these models are usually written in terms of aerodynamic derivatives.

The state and control matrices being respectively A and B, a steady state X_e is locally stable if the real parts of all the eigenvalues of the A matrix are negative. If the real part of any eigenvalue of the matrix A is positive, the steady state is locally unstable. The system is attracted to a stable steady state in its neighborhood and repelled if the steady state is unstable. The eigenvectors of the state matrix A characterize subspaces of different modes of disturbed motion. The linearized response to control input is characterized by the control matrix B. The state matrix A and control matrix B are calculated on a grid of points for different flight regimes and steady maneuvers provide a linear parameter varying approximation for the original nonlinear system which is equally important for the open loop dynamic analysis and for control law design.

5.2.2.1 Properties of the Longitudinal Linear Model

The state equation is solved to obtain the response transfer functions. Since the solution involves algebraic manipulation of matrices it is necessary to first obtain the Laplace transform of the state equation thus, assuming zero initial conditions,

$$X(s) = \left(sI_{4\times4} - A_{long}\right)^{-1} B_{long} U(s) \tag{5.17}$$

where $I_{4\times4}$ is the 4×4 unit matrix

Typically, the longitudinal characteristic equation has two real and one complex pair of roots, each root describing a stability mode. Analysis of the eigenvalues of the state matrix A enables an approximate description of the stability modes to be made [48, 108, 201]

- The surge mode is described by the biggest real root of $\Delta(s)$. The mode appears as an exponential speed subsidence usually the mode is stable with a long time constant. It is associated to the forward speed $\bar{u}_e$
- The heave mode is described at the hover by the second real root of $\Delta(s)$. It is associated with the vertical speed $\bar{w}_e$ or equivalently the angle of attack α.
- The longitudinal pendulum mode corresponds to the complex pair of poles that is associated to the pitch angle $\bar{\theta}_e$ and the pitch rate $\bar{q}_e$.

5.2.2.2 Properties of the Lateral Linear Model

Typically the lateral characteristic equation also has two real roots and one complex pair of roots in its solution.

- The fast mode: Yaw subsidence mode is related to the yaw rate $\bar{r}$ and presents a time constant that decreases with the airspeed. The mode appears as an exponential speed subsidence usually the mode is stable with a long time constant.
- The slow mode, usually named, side slip subsidence mode is associated with the lateral speed $\bar{v}_e$ or equivalently the angle of side slip β.
- In the hover condition, the zero damping of the complex pair characterizes the oscillatory roll mode related to the roll rate $\bar{p}_e$ and the roll angle $\bar{\phi}_e$. The oscillatory rolling movement is the lateral equivalent of the longitudinal pendulum oscillation and arises because the center of gravity is located below the center of lift of the lighter than air robot. With the increase of airspeed, the general stability improves since the damping ratio of the oscillatory mode increases and the eigenvalues of the real modes become more negative.

5.2.2.3 Performance Maneuverability

The evaluation of the lighter than air robot flight performance and maneuvering capabilities is an important task of development. Flight performance is an important task of development and reflects a lighter than air robot's ability to perform steady coordinated maneuvers at different speeds and altitudes (e.g. its capability of maintaining a straight and level flight, a steady level turn, climbing or gliding turns). The maneuverability characterizes an aerial vehicle ability to alter its steady flight trajectory via rotation with respect to the flight velocity vector. The level of maneuverability is directly linked with attainable values for the angle of attack or the normal load factor and the angular rate in the velocity vector roll maneuver.

The steady performance and maneuvering capabilities are usually evaluated by solving the steady-state problem for the rigid-body equations of motion, which depend on the propulsion, aerodynamics mass and inertia characteristics. For example, the steady states can be determined for any particular combination of control inputs without setting any requirement for maneuver parameters. Alternatively, some steady state and control variables can be chosen as independent and the remaining variables can be determined according to the imposed kinematical constraints.

The continuation and bifurcation analysis method have evolved into a powerful tool for lighter than air robot trim and stability analysis. Within this computational framework, the equilibrium states are computed in the extended space of state variables and one selected control parameter, whereas all other control effectors are kept constant [43, 48, 49, 55, 181].

The steady states are represented as a one parameter continuation diagram or as a two-dimensional equilibrium surface. At some ranges of control parameters, folded dependencies may appear in the steady state diagrams and surfaces, which indicate potential nonlocal departures in motion variables at limit points of the folds. Other types of bifurcations or changes in steady state local stability are also identified during the continuation procedure, thus allowing effective prediction of lighter than air robot instability and loss of control. Imposing additional maneuver specific kinematical constraints, allows the continuation of steady states representing a specified maneuver. In this constrained problem, several control surfaces are deflected for designing airship steady state maneuvers.

The computation problem for steady states is formulated for equations of motions, augmented by auxiliary equations specifying maneuver kinematics and maneuver parameters. The solutions of this constrained trim problem are presented in the form of attainable equilibrium sets computed on a grid of points in the plane of two selected parameters characterizing the flight regime and steady maneuver. All steady states are classified according to the eigenvalue spectrum of the linearized system. This classification generates local stability maps, giving a qualitative insight into the lighter than air robot nonlinear dynamics. Deflection constraints on the lighter than air robot control inputs limit the available control power so that the set of equilibrium states for the nonlinear system is bounded.

5.2.3 Classical Linear Control

For the design of the linear controllers, several control design methodologies are available within control theory [56, 57, 70, 108, 119, 150]. Pole Placement (PP) and Linear Quadratic Regulator (LQR) are among the most popular modern controller techniques for Multiple Input Multiple Output system (MIMO). Pole placement allows to allocate the poles of the MIMO system to desired locations in one step by solving equations for the feedback gains or by using the Bass-Gura approach. The pole placement strategy does not confer any stability robustness to the closed loop system. This is an important factor since any model of the system is an approximation of the real nonlinear dynamics. Furthermore, these models do not take into account disturbances such as wind gusts or sensor measurement noise. Robustness to model parameter errors and to disturbances is, in fact, a key issue in the choice of the controller.

Optimal control methods are attractive because they handle multi-input-multi-output systems easily and aid in the selection of the desired pole location. The Linear Quadratic Regulator is the solution to an optimization problem that has some attractive properties. The optimal controller automatically ensures a stable closed loop system, achieves guaranteed levels of stability robustness and is simple to compute. This control minimizes a quadratic cost functional such that constraints imposed by the system dynamics are verified. The design weights (state and control weighting matrices) are the tools to balance the state errors against the control effort. In the lighter than air robot control case, the control weighting matrix is a specially important tool in the sense that it allows the designer to change the control effort of the different actuators over the flight envelope.

Remark 5.2 Gain scheduling is the prevailing flight control design methodology while the linear control is only valid around a single equilibrium condition, the gain scheduling solution performs for a large set of trim conditions and then constructs a gain schedule by considering gains with respect to flight conditions.

5.2.3.1 Pole Placement by Feedback

Pole Placement by State Feedback The key idea for the control-pole selection is that the poles are chosen so that the design specifications are met while the use of control is kept to a level needed to meet the specifications. This pole selection criterion keeps the actuator sizes to a minimum, which helps to minimize the cost and weight of the control system. Consider a linear time invariant dynamic system described by:

$$\dot{X} = AX + BU \tag{5.18}$$

where $X \in R^n, U \in R^m$. In state feedback control, the state vector is multiplied by a gain matrix K and feedback into the control input:

$$U = -KX \tag{5.19}$$

where $K \in R^{m \times n}$. The closed loop system is then described by:

$$\dot{X} = (A - BK)X \tag{5.20}$$

and the closed loop characteristic equation becomes

$$|sI - A + BK| = 0 \tag{5.21}$$

If the system is controllable, the eigenvalues of the closed loop system can be arbitrarily assigned, provided that the real parts are negative and the complex conjugate eigenvalues appear in pairs.

Pole Placement by Output Feedback In feedback control the value of the control input is chosen not as an explicit function of time, but on the basis of an observed output. Unexpected events, small disturbances or miscalculations due to uncertain parameters can be taken into consideration by feedback control but not by open loop control.

$$\begin{aligned} \dot{X} &= AX + BU \\ Y &= CX \end{aligned} \tag{5.22}$$

A control law is thus defined by

$$U = KY \tag{5.23}$$

where the matrix K is called the feedback gain matrix. This leads to the equation

$$\dot{Y} = \left(CAC^{-1} + CBK\right)Y \tag{5.24}$$

It comes down to choosing, for given matrices A, B, C, the feedback matrix such that the pair $[(CAC^{-1} + CBK), K]$ has desirable properties.

The poles of the closed loop system are very important features for judging the behavior of the closed loop system. When all poles lie in the open left half plane, the system is asymptotically stable. The pole placement problem concerns the choice of the feedback gain matrix K such that specified closed loop locations are achievable. The closed loop poles or equivalently the closed loop characteristic polynomial can be chosen if and only if the system is controllable.

$$\text{rank}\left[B, AB, \ldots, A^{n-1}B\right] = n$$

An equilibrium point of a nonlinear system can be stabilized if the linearized system is controllable, or more generally, stabilizable.

5.2.3.2 Linear Quadratic Regulator and Estimator

Linear Quadratic Regulator The gain matrix K of the state-feedback $U = -KX$ can be determined by minimizing the linear quadratic performance index

$$J = \frac{1}{2}\int_0^T \left(X^T Q_1 X + U^T R_1 U\right) dt \tag{5.25}$$

where Q_1 is the state weighting matrix and R_1 is the control input weighting matrix. The gain matrix K is then obtained as

$$K = R_1^{-1} B^T S \tag{5.26}$$

by solving the algebraic Riccati equation

$$\frac{dS}{dt} = A^T S + SA - SBR_1^{-1}B^T S + Q_1 \tag{5.27}$$

Certain conditions must be met for a unique positive definite solution to the above Riccati equation to exist:

1. The matrix Q_1 must symmetric and positive semi-definite i.e. $Q_1 = Q_1^T \geq 0$
2. The matrix R_1 must be symmetric positive definite i.e. $R_1 = R_1^T > 0$
3. The pair (A, B) must be controllable (stabilizable)
4. The pair (A, H) must be observable (detectable) where $H^T H = Q_1$ and $\text{rank}\, H = \text{rank}\, Q_1$

When $T \to \infty$ then the Riccati equation becomes

$$0 = A^T S + SA - SBR^{-1}B^T S + Q_1 \tag{5.28}$$

Linear Quadratic Estimator To determine the estimator gain matrix **L** using the Linear Quadratic Estimator (LQE), the system described by the following state space equations is considered:

$$\begin{aligned} \dot{X} &= AX + BU + GW_N \\ Y &= CX + V_N \end{aligned} \tag{5.29}$$

where W_N is the process noise and V_N is the measurement noise, both W_N, V_N are assumed to be white noise processes with

$$\begin{aligned} \mathbf{E}\left[W_N(t)W_N^T(\tau)\right] &= \mathbf{W}_N\delta(t-\tau) \\ \mathbf{E}\left[V_N(t)V_N^T(\tau)\right] &= \mathbf{V}_N\delta(t-\tau) \end{aligned} \tag{5.30}$$

where $\mathbf{W}_N$ and $\mathbf{V}_N$ are the corresponding spectral density matrices. The gain matrix $\mathbf{L}$ of the linear quadratic estimator is then selected such that the observation error:

$$e = X - \hat{X}$$

is minimized in the presence of noise, by solving the algebraic Riccati equation

$$AS + SA^T - SC^T\mathbf{V}_N^{-1}CS + G\mathbf{W}_N G^T = 0 \tag{5.31}$$

where S is the estimate error covariance matrix and

$$L = SC^T\mathbf{V}_N^{-1}$$

The linear quadratic Gaussian (LQG)/linear quadratic estimator technique offers an optimal compensator design in the presence of random disturbances given certain weighting parameters for the states and the control inputs and certain parameters describing the random disturbances. The question remains of how to choose these parameters and what choice provides the best optimal design.

5.2.4 Linear Robust Control

At the heart of robust control is the concept of an uncertain Linear Time Invariant (LTI) model. Model uncertainty arises when system gains or other parameters are not precisely known or can vary over a given range. To be robust, the control system must meet given stability and performance requirements for all possible values of uncertain parameters.

Some of the parameters that describe the lighter than air robot are likely to be uncertain. These parameters are mostly the aerodynamic model parameters, obtained in wind tunnel experiments. The weighting mass or heaviness, which represents the difference between the weight and buoyancy forces, is also considered, since the equilibrium flight is mostly affected by its value. The control laws are designed considering a deterministic model of the lighter than air robot named nominal. However, the real system has a wind disturbance input, since in a real flight, wind disturbances are always present. Complex models are not always required for good control. However, optimization methods (including methods based on H_∞, H_2 and μ synthesis optimal control theory) generally tend to produce controllers with at least as many states as the plant model. Unlike traditional optimal control, robust optimal control minimizes the influence of various types of uncertainties in addition to (or even instead of) performance and control energy optimization. Generally, this implies design of a possibly low gain controller with reduced sensitivity to input changes. As a consequence, robust controllers often tend to be conservative and slow. On the other hand, they may be thought as the stabilizing controllers for the whole set of plants (which include a range of uncertainties) and not only for the modeled system. The majority of robust control techniques have origins in the classical frequency domain methods. The key modification of the classic methods is shifting from eigenvalues to singular values (of the transfer function that describes the system), the singular value Bode plot being the major indicator of multi-variable feedback system performance. Probably the most popular modern robust control design techniques (particularly in the aerospace field) are H_∞, H_2 control, also known as the frequency-weighted LQG synthesis and the small gain problem respectively. These techniques and the underlying theories are described in several works [33, 57, 88, 119, 171, 193, 230].

Remark 5.3 The H_∞ control design found extensive use for aircraft flight control. One of the first such applications was the development of controllers for the longitudinal control of a Harrier jump jet [88]. This work has been extended in [171] to fully integrated longitudinal, lateral and propulsive control. Other works include [98], where a lateral autopilot for a large civil aircraft is designed, and [106], where a robust longitudinal controller subject to aircraft weight and center of gravity uncertainty is demonstrated. A mixed H_∞, H_2 approach is applied in [193] to design an auto land controller for a large commercial aircraft. The method employed here utilizes the H_2 controller for slow trajectory tracking and the H_∞ controller for fast dynamic robustness and disturbance rejection. Several H_∞ controllers have been tried to accomplish the UAV shipboard launch task [50]. It has been found that these controllers perform quite well in nominal situations. However, in the presence of large disturbances which place the aircraft well beyond its linear design operating point, the controllers performed poorly (sometimes extremely). At the same time, inability to include even simple static nonlinearities such as time delays and saturations made it difficult to synthesize a practical controller for this task within linear approach. Another deficiency found is common to all frequency domain techniques: the frequency domain performance specifications cannot be rigidly translated into time and spatial domain specifications.

The time domain performance can be accounted for directly in the time domain l_1 design [33]. It is often extended to include the frequency domain objectives, resulting in a mixed norm approach. However, l_1 design is plagued by the excessive order of the generated controllers. This is usually solved by reducing the problem to suboptimal control, imposing several restrictions on the system and performance specifications. Controllers with excessive order will generally be produced when using practical constraints. There have been attempts to solve the H_∞ optimal control problems for nonlinear systems. However, these methods usually rely on very limiting assumptions about the model, uncertainties and disturbance structure. One mathematical development of nonlinear H_∞ control can be found in [193, 218, 230].

5.2.4.1 Modeling Errors and Stability Robustness

In the design of a lighter than air robot control system, it is important to realize that the rigid body equations are only an approximation to the nonlinear dynamics. A lighter than air robot has flexible modes that are important [10]. These unmodeled dynamics can act to destabilize, a control system that may have quite suitable behavior in terms only of rigid body model.

Moreover, as the lighter than air robot changes its equilibrium flight condition, the linearized rigid body model describing its perturbed behavior changes. This parameter variation is a low frequency effect that can also act to destabilize the system. To compensate for this variation, one may determine suitable controller gains for linearized models at several design equilibrium points over a flight envelope. Then, these design gains may be scheduled in computer look up tables for suitable controller performance over the whole envelope.

Gain scheduling is a process by which one of several different control algorithms is chosen based on some operating conditions and is thus adaptive. Since the control algorithms are designed off-line with a priori-information, gain scheduling algorithm must identify the proper control design to another during system operation. Effectively, the design is broken into regions of operation and in each, a fixed control design is used.

The control parameters are changed in small discrete amounts at a rate much slower than the slowest time constant of the closed loop system, the system stability can be ensured by examining each of the control laws as if it were fixed. Thus gain scheduling is usually defined as dependent on some slowly varying parameter (relative to the control bound width). Although gain scheduling is adaptive, it is perhaps the least sophisticated of the strategies of adaptive control. This is because a relatively number of designs are done off-line and loaded into the controller as options that are subsequently chosen during operation as a function of some measured or computed conditions. It is the predominant method of control used to handle the wide plant variations that occur in flight control system.

Remark 5.4 For gain scheduling to work, it is essential for the controller gains at each design equation point to guarantee stability for actual flight conditions near that equilibrium point. Thus, it is important to design controllers that have stability robustness in spite of modeling errors due to high frequency unmodeled dynamics and plant parameter variations.

It is often important to account for disturbances such as wind gusts and also for sensor measurement noise. Disturbances can often act to cause unsatisfactory performance in a system that has been designed without taking them into account. Thus it is important to design controllers that have performance robustness, which is the ability to guarantee acceptable performance (in terms for instance of percent overshoot, settling time, ...) even though the system may be subject to disturbances.

For robust performance, the minimum singular value of the closed loop gain should be large at low frequencies, where disturbances are present. On the other hand, for robust stability the maximum singular value of the loop gain should be small at high frequencies, where there are significant modeling inaccuracies. To guarantee stability despite parameter variations in the linearized model due to operating point changes, the maximum singular value should be below an upper limit. Inputs in a certain direction in the input space will excite only the singular values associated with that direction.

It is important for the control gains to stabilize the lighter than air robot at all points near the design operating point for this gain scheduling procedure to be effective. In passing from operating point to operating point, the parameters of this state variable model vary. Controllers that guarantee robust stability despite plant parameter variations must be designed. Suppose the nominal perturbed model is

$$\begin{aligned} \dot{X} &= AX + BU \\ Y &= CX \end{aligned} \tag{5.32}$$

However, due to operating changes, the actual airship perturbed motion is described by:

$$\begin{aligned} \dot{X} &= (A + \Delta A)\,X + (B + \Delta B)\,U \\ Y &= (C + \Delta C)\,X \end{aligned} \tag{5.33}$$

Where the plant parameter variation matrices are $\Delta A, \Delta B, \Delta C$. The transfer function for this representation is given as:

$$G'(s) = G(s) + \Delta G(s) \tag{5.34}$$

with

$$G(s) = C\,(sI - A)^{-1}\,B \tag{5.35}$$

and

$$\begin{aligned} \Delta G(s) = {} & C\,(sI - A)^{-1}\,\Delta B + \Delta C\,(sI - A)^{-1}\,B \\ & + C\,(sI - A)^{-1}\,\Delta A\,(sI - A)^{-1}\,B \end{aligned} \tag{5.36}$$

where second order effects have been neglected

To incorporate the robustness concept into the Linear Quadratic output feedback design procedure, the following steps may be accomplished:

1. If necessary, augment the system with added dynamics to achieve the required steady state error behavior or to achieve balanced singular value at closed loop.
2. Select a performance index, the Performance Index weighting matrices Q_1 and R_1
3. Determine the optimal output feedback gain K
4. Simulate the time responses of the closed loop system to verify that they are satisfactory. If not, select different Q, R, K and return to step 3

5. Determine the low frequency and high frequency bounds required for performance robustness and stability robustness. Plot the loop gain singular values to verify that the bounds are satisfied. If they are not, select different Q_1 and R_1, K and return to step 3

5.2.4.2 Standard H_∞

Consider a linear time varying invariant system described by:

$$\begin{aligned}\dot{X}(t) &= AX(t) + B_1 d(t) + B_2 U(t)\\ Z(t) &= C_1 X(t) + D_{11} d(t) + D_{12} U(t)\\ Y(t) &= C_2 X(t) + D_{21} d(t) + D_{22} U(t)\end{aligned} \tag{5.37}$$

where $X \in R^n$, $d \in R^{m_1}$, $U \in R^{m_2}$, $Y \in R^{p_1}$, $Z \in R^{p_2}$ are respectively the state, disturbance input, control input, controlled output and measured output vectors.

The transfer function representation of this system is given by:

$$\begin{aligned}\begin{pmatrix} Z(s)\\ Y(s)\end{pmatrix} &= \begin{pmatrix} P_{11}(s) & P_{12}(s)\\ P_{21}(s) & P_{22}(s)\end{pmatrix}\begin{pmatrix} d(s)\\ U(s)\end{pmatrix}\\ &= \left\{\begin{pmatrix} C_1\\ C_2\end{pmatrix}(sI-A)^{-1}(B_1, B_2) + \begin{pmatrix} D_{11}(s) & D_{12}(s)\\ D_{21}(s) & D_{22}(s)\end{pmatrix}\right\}\begin{pmatrix} d(s)\\ U(s)\end{pmatrix}\end{aligned} \tag{5.38}$$

where $P_{ij}(s)$ are real-rational transfer function matrices and $P(s) = (P_{ij}(s), i = 1, 2; j = 1, 2)$ is called the generalized plant, which may include the internal feedback loop model and frequency-dependent weightings.

For a linear system with a feedback control of the form

$$U(s) = -K(s)Y(s) \tag{5.39}$$

The closed-loop transfer function from D to Z can be derived as:

$$T_{zd}(P, K) = P_{11} + P_{12}K\left(I - P_{12}K\right)^{-1} P_{21} \tag{5.40}$$

where $K(s)$ is a compensator transfer function matrix to be synthesized. The H_∞ norm of a real rational transfer function matrix $T(s)$ is defined as

$$\|T(s)\|_\infty = \sup_\omega \bar{\sigma}\left(T(j\omega)\right) \tag{5.41}$$

where $\bar{\sigma}(T(j\omega))$ denotes the largest singular value of $T(j\omega)$ for a given ω. The H_∞ space consists of functions that are stable and bounded.

The design objective of standard H_∞ control problem is then to find $K(s)$ subject to $\|T_{zd}(P, K)\|_\infty < \gamma$ for some specified $\gamma \in R$.

Using the input-output decomposition of structured plant parameter variations in terms of the fictitious inputs and outputs of an internal uncertainty loop to be discussed next, the state space solution to the standard H_∞ control problem will be utilized to design parameter-insensitive parameters [218].

Modeling of Structured Parameter Uncertainty Consider an uncertain linear dynamic system described by:

$$E\dot{X} = FX + G_d d + G_u U \tag{5.42}$$

where X, d, U are the state, external disturbance and control input vectors, respectively, G_d is the disturbance distribution matrix, G_u is the control input distribution matrix and the matrices E, F are subject to structured parameter variations.

Suppose that there are l independent uncertain parameter variables δ_i and assume that the perturbed matrices E, F can be linearly decomposed as follows:

$$\begin{aligned} E &= E_0 + \Delta E \\ F &= F_0 + \Delta F \end{aligned} \tag{5.43}$$

where E_0, F_0 are the nominal matrices and $\Delta E, \Delta F$ are the perturbation matrices defined as:

$$\begin{aligned} \Delta E &= \sum_{i=1}^{l} \Delta E_i \delta_i = \sum_{i=1}^{l} M_E^i \delta_i I_{K_i} N_E^i = M_E \varepsilon_E N_E \\ \Delta F &= \sum_{i=1}^{l} \Delta F_i \delta_i = \sum_{i=1}^{l} M_F^i \delta_i I_{\nu_i} N_F^i = M_F \varepsilon_F N_F \end{aligned} \tag{5.44}$$

where K_i is the rank of ΔE_i, ν_i is the rank of ΔF_i and $\varepsilon_E, \varepsilon_F$ are diagonal matrices with δ_i as their diagonal elements.

If $K_i = \nu_i = 1$ for $i = 1, \ldots, l$ a special case of rank-one dependency, $M_E^{(i)}, M_F^{(i)}$ become column vectors and $N_E^{(i)}, N_F^{(i)}$ become row vectors. In this case, there is no repeated elements δ_i in $\varepsilon_E, \varepsilon_F$.

Let $\varepsilon = \text{diag}(\varepsilon_E, \varepsilon_F)$,

$$\tilde{z} = \begin{pmatrix} \tilde{z}_E & \tilde{z}_F \end{pmatrix} = \begin{pmatrix} N_E \dot{x}_E & N_F \dot{x}_F \end{pmatrix}; \qquad \tilde{d} = -\varepsilon \tilde{z}$$

where d is called the fictitious disturbance input, $\tilde{z}$ the fictitious output and ε the gain matrix of a fictitious internal uncertainty loop, which is caused by uncertainty in the matrices E, F, then substituting, we obtain

$$E_0 \dot{X} = F_0 X + G_{\tilde{d}} \tilde{d} + G_d d + G_u U \tag{5.45}$$

where $G_{\tilde{d}} = (M_E; M_F)$. Defining the controlled output vector as:

$$z = \begin{pmatrix} C_{11} \\ 0 \end{pmatrix} X + \begin{pmatrix} 0 \\ I \end{pmatrix} U \tag{5.46}$$

and introducing new variables

$$\hat{d} = \begin{pmatrix} \tilde{d} \\ d \end{pmatrix}; \qquad \hat{z} = \begin{pmatrix} \tilde{z} \\ z \end{pmatrix} \tag{5.47}$$

We obtain a modified state space representation of the system as follows

$$\dot{X} = AX + B_1\hat{d} + B_2U$$
$$\hat{z} = C_1X + D_{11}\hat{d} + D_{12} \tag{5.48}$$

where

$$A = E_0^{-1}F_0; \qquad B_1 = E_0^{-1}\left(G_{\tilde{d}} \quad G_d\right); \qquad B_2 = E_0^{-1}G_u$$

Note that $D_{11} = 0$ if there is no uncertainty in E.

$$D_{11} = N_E E_0^{-1}\begin{pmatrix} G_{\tilde{d}} & G_d \\ 0 & 0 \\ 0 & 0 \\ 0 & 0 \end{pmatrix} \qquad D_{12} = \begin{pmatrix} N_E E_0^{-1}G_u \\ 0 \\ 0 \\ 1 \end{pmatrix} \qquad C_1 = \begin{pmatrix} N_E E_0^{-1}F_0 \\ N_F \\ C_{11} \\ 0 \end{pmatrix}$$

Robust H_∞ Compensator Design Consider an uncertain linear system described by

$$\begin{pmatrix} \hat{Z}(s) \\ Y(s) \end{pmatrix} = \begin{pmatrix} P_{11}(s) & P_{12}(s) \\ P_{21}(s) & P_{22}(s) \end{pmatrix}\begin{pmatrix} \hat{d}(s) \\ U(s) \end{pmatrix}$$
$$= \left\{\begin{pmatrix} C_1 \\ C_2 \end{pmatrix}(sI - A)^{-1}(B_1, B_2) + \begin{pmatrix} D_{11}(s) & D_{12}(s) \\ D_{21}(s) & D_{22}(s) \end{pmatrix}\right\}\begin{pmatrix} \hat{d}(s) \\ U(s) \end{pmatrix} \tag{5.49}$$

where $X \in R^n, \hat{d} \in R^m, U \in R_{m_2}, \hat{z} \in R^{p_1}, y \in R_{p_2}$ are respectively the state, augmented disturbance input, control input vectors, augmented controlled output and measured output vectors. Furthermore, the fictitious disturbance input $\tilde{d}$ and the fictitious output $\tilde{z}$ are defined as

$$\tilde{d} = \Delta\tilde{z}$$

where Δ is the gain matrix of the internal uncertainty loop. This uncertain system to be controlled can also be described by

$$\begin{pmatrix} \tilde{z} \\ z \\ y \end{pmatrix} = \begin{pmatrix} G_{11} & G_{12} & G_{13} \\ G_{21} & G_{22} & G_{23} \\ G_{31} & G_{32} & G_{33} \end{pmatrix}\begin{pmatrix} \tilde{d} \\ d \\ u \end{pmatrix} \qquad \tilde{d} = \Delta\tilde{z} \qquad U = -K(s)Y \tag{5.50}$$

where $K(s)$ is a feedback compensator to be designed.

After closing the control loop with a stabilizing controller $K(s)$ but with the internal uncertainty loop broken, the following representation of the closed loop system is obtained.

$$\hat{z} = T_{\hat{z}\hat{d}}\hat{d} \tag{5.51}$$

where

$$T_{\hat{z}\hat{d}} = \begin{pmatrix} T_{11} & T_{12} \\ T_{21} & T_{22} \end{pmatrix}$$

$$T_{11} = G_{11} - G_{13}K\,(I + G_{33}K)^{-1}\,G_{31}$$

$$T_{12} = G_{12} - G_{13}K\,(I + G_{33}K)^{-1}\,G_{32}$$

$$T_{21} = G_{21} - G_{23}K\,(I + G_{33}K)^{-1}\,G_{31}$$

$$T_{22} = G_{22} - G_{23}K\,(I + G_{33}K)^{-1}\,G_{32}$$

The actual closed-loop transfer function matrix from d to z under plant perturbations becomes

$$T_{zd} = T_{22} + T_{21}\Delta\,(I - T_{11}\Delta)^{-1}\,T_{12} \tag{5.52}$$

The following propositions provide sufficient conditions for stability performance robustness [46, 218].

Proposition 5.1 (Stability robustness) *If* $\|T_{11}(s)\| < \infty$ *then* $T_{zd}(s, \alpha_1\Delta), \forall d \in [0, 1]$ *is stable for* $\|\Delta\| \leq \gamma_1^{-1}$.

Proposition 5.2 (Performance robustness) *If* $\|T_{\hat{z}\hat{d}}(s)\| < \gamma_1$ *then* $T_{zd}(s, \alpha_1\Delta)\ \forall\alpha_1 \in [0, 1]$ *is stable and* $\|T_{zd}(s, \alpha_1\Delta)\|_\infty < \gamma_1,\ \forall d \in [0, 1]$ *with* $\|\Delta\| \leq \gamma_1^{-1}$.

Because T_{11} can be represented as

$$T_{11} = \begin{bmatrix} I & 0 \end{bmatrix} \begin{pmatrix} T_{11} & T_{12} \\ T_{21} & T_{22} \end{pmatrix} \begin{pmatrix} 0 \\ I \end{pmatrix}$$

so it can be concluded that $\|T_{11}\|_\infty \leq \|T_{\hat{z}\hat{d}}(s)\|$. Consequently, if the condition $\|T_{\hat{z}\hat{d}}(s)\| < \gamma_1$ both stability and performance robustness will be achieved with respect to bounded uncertainty $\|\Delta\| \leq \gamma_1^{-1}$.

Proposition 5.3 (H_∞ Suboptimal controller) *Consider a linear system described by*:

$$\begin{aligned} \begin{pmatrix} \hat{z}(s) \\ y(s) \end{pmatrix} &= \left\{ \begin{pmatrix} C_1 \\ C_2 \end{pmatrix} (sI - A)^{-1} \begin{pmatrix} B_1 & B_2 \end{pmatrix} + \begin{pmatrix} D_{11} & D_{12} \\ D_{21} & D_{22} \end{pmatrix} \right\} \begin{pmatrix} \hat{d}(s) \\ u(s) \end{pmatrix} \\ &= \begin{pmatrix} P_{11}(s) & P_{12}(s) \\ P_{21}(s) & P_{22}(s) \end{pmatrix} \begin{pmatrix} \hat{d}(s) \\ d(s) \end{pmatrix} \end{aligned} \tag{5.53}$$

Assume the following

1. (A, B_2) *is stabilizable and* (C_2, A) *is detectable*
2. $D_{12}^T(C_1 D_{12}) = (0\ I)$
3. $\begin{pmatrix} B_1 \\ D_{21} \end{pmatrix} D_{21}^T = \begin{pmatrix} 0 \\ I \end{pmatrix}$
4. *The rank of* $P_{21}(j\omega)$ *and* $P_{12}(j\omega)$ *is* p_2 *and* m_2, *respectively for all* ω
5. $D_{11} = 0$ *and* $D_{22} = 0$

There exists an internally stabilizing controller such that $\|T_{\hat{z}\hat{d}}\| < \infty$ *if and only if the following Riccati equations*

$$0 = A^T S + SA - S\left(B_2 B_2^T - \gamma_1^{-2} B_1 B_1^T\right) S + C_1^T C_1$$

$$0 = AS' + S'A^T - S'\left(C_2^T C_2 - \gamma_1^{-2} C_1^T C_1\right) S' + B_1 B_1^T$$

have solutions S *and* S'. *An* H_∞ *suboptimal controller that satisfies* $\|T_{\hat{z}\hat{d}}\| < \infty$ *where* γ_1 *is a design tradeoff variable specifying an upper bound of the perturbed closed loop transfer matrix* $T_{\hat{z}\hat{d}}$ *is then obtained as*:

$$\dot{\hat{x}} = A_c \hat{x} + Ly$$

$$u = K\hat{x}$$

where

$$K = B_2^T S$$

$$L = \left(I - \gamma_1^{-2} S' S\right)^{-1} S' C_2^T$$

$$A_c = A + \gamma_1^{-2} B_1 B_1^T S - B_2 K - LC_2$$

and $\hat{x}$ *represents the controller state vector.*

The closed loop system (neglecting all of the external inputs) is then described as:

$$\begin{pmatrix} \dot{x} \\ \dot{\hat{x}} \end{pmatrix} = \begin{pmatrix} A & -B_2 K \\ LC_2 & A_c \end{pmatrix} \begin{pmatrix} x \\ \hat{x} \end{pmatrix} \tag{5.54}$$

This controller has a structure similar to a conventional state space controller, consisting of an estimator and a regulator, but is designed for a plant system matrix

$$A + \gamma_1^{-2} B_1 B_1^T S$$

Consequently, the separation principle of the conventional LQG technique does not hold here.

5.3 Nonlinear Control

The designer of a control system must define basic things:

- A model of the plant to be controlled and the range of its validity
- The nominal value of the model parameters and their expected deviation

- The performance objectives
- The constraints on the design, such as the cost of control action, control authority limits and the intended cost of the controller

Design of controls for nonlinear systems may be placed in several categories: the most primitive is feedback linearization. Parameter variations of the lighter than air robot models can have severe impact on performance and stability.

5.3.1 Dynamic Inversion

Some research has been done for the control systems design for maneuvering the lighter than air robot. Dynamic inversion (or feedback linearization) is a methodology to design closed loop control laws for non linear systems. It searches for a global control law from a single global nonlinear model of the plant. A fundamental assumption in the Dynamic inversion methodology is that the plant dynamics should be perfectly modeled and may be canceled exactly. In practice, this assumption is not realistic, and the robustness of the closed loop dynamics must be secured, in order to suppress any undesired behavior due to plant uncertainties. An outer loop controller can be used to improve a Dynamic inversion inner loop controller robustness to uncertainties as wind and turbulence disturbances as well as uncertainties in the model parameters [195, 198].

5.3.1.1 Dynamic Inversion Approach

Dynamic inversion is a controller design technique by which existing undesirable dynamics are canceled out and replaced by designer specified appropriate ones.

Consider the nonlinear system affine in control given by:

$$\begin{aligned} \dot{X}_1 &= B_1 X_2 \\ \dot{X}_2 &= f_0(X_2) + f_1(X_4) \\ \dot{X}_3 &= (B_2 + B_4)\left(f_0(X_2) + f_1(X_4)\right) + B_3 U \\ \dot{X}_4 &= U \\ Y &= X_1 \end{aligned} \tag{5.55}$$

where the state vector is given by $X = (X_1, X_2, X_3, X_4)$, $\dim(X) = 12$ where

$$X_1 = \begin{pmatrix} x \\ y \\ z \end{pmatrix} \quad X_2 = \begin{pmatrix} V \\ \chi \\ \gamma \end{pmatrix} \quad X_3 = \begin{pmatrix} \dot{V} \\ \dot{\chi} \\ \dot{\gamma} \end{pmatrix} \quad X_4 = \begin{pmatrix} T \\ \sigma \\ \alpha \end{pmatrix}$$

The control inputs are

$$U = \begin{pmatrix} \dot{T} \\ \dot{\sigma} \\ \dot{\alpha} \end{pmatrix} \quad \text{and} \quad Y = X_1$$

contains the output variables to be controlled. f_0, f_1 are non linear functions given respectively by:

$$f_0(X_2) = \begin{pmatrix} \frac{(B-mg)\sin\gamma}{m+m_{11}} \\ 0 \\ -\frac{(B-mg)\cos\gamma}{(m+m_{33})V} \end{pmatrix} \qquad f_1(X_4) = \begin{pmatrix} \frac{(T\cos\alpha - D)}{m+m_{11}} \\ \frac{(T\sin\alpha+L)\sin\sigma}{(m+m_{22})V\cos\gamma} \\ -\frac{((T\sin\alpha+L)\cos\sigma)}{(m+m_{33})V} \end{pmatrix}$$

The different matrices are given as

$$B_1 = \begin{pmatrix} \cos\gamma\sin\chi & 0 & 0 \\ \cos\gamma\cos\chi & 0 & 0 \\ \sin\gamma & 0 & 0 \end{pmatrix}$$

$$B_2 = \begin{pmatrix} 0 & 0 & \frac{(B-mg)\cos\gamma}{m+m_{11}} \\ 0 & 0 & 0 \\ -\frac{(B-mg)\cos\gamma}{(m+m_{33})V^2} & 0 & \frac{(B-mg)\sin\gamma}{(m+m_{33})V} \end{pmatrix}$$

$$B_3 = \begin{pmatrix} -\frac{\cos\alpha}{m+m_{11}} & 0 & -\frac{T\sin\alpha - D_\alpha}{m+m_{11}} \\ \frac{\sin\alpha}{(m+m_{22})V\cos\gamma} & \frac{(T\sin\alpha+L)\cos\sigma}{(m+m_{22})V\cos\gamma} & \frac{(T\cos\alpha+L_\alpha)\sin\sigma}{(m+m_{22})V\cos\gamma} \\ \frac{\sin\alpha}{(m+m_{33})V} & -\frac{(T\sin\alpha+L)\sin\sigma}{(m+m_{33})V} & \frac{(T\cos\alpha+L_\alpha)\cos\sigma}{(m+m_{33})V} \end{pmatrix}$$

$$B_4 = \begin{pmatrix} -\frac{D_v}{m+m_{11}} & 0 & 0 \\ -\frac{L_v\sin\sigma}{(m+m_{22})V\cos\gamma} + \frac{(T\sin\alpha+L)\sin\sigma}{(m+m_{22})V^2\cos\gamma} & 0 & \frac{(T\sin\alpha+L)\sin\sigma V\cos\sigma}{(m+m_{22})V^2\cos^2\gamma} \\ -\frac{(T\sin\alpha+L)\cos\sigma}{(m+m_{33})V^2} & 0 & 0 \end{pmatrix}$$

with $D_\alpha = \frac{\partial D}{\partial \alpha}$, $L_\alpha = \frac{\partial L}{\partial \alpha}$, $D_v = \frac{\partial D}{\partial V}$, $L_v = \frac{\partial L}{\partial V}$.

In order to deduce the input necessary to track a desired output Y_{ref}, Y is derived twice, yielding

$$\ddot{Y} = H_v\left(f(V,P) + BU\right) + H_p h(V,P) \tag{5.56}$$

where H_v and H_p matrices are the partial derivatives of h with respect to V and P respectively. The inversion of the output dynamic equation is thus given by

$$U = (H_v B)^{-1}\left(\ddot{Y} - \left(H_v f(V,P) + H_p h(V,P)\right)\right) \tag{5.57}$$

which state the necessary input value U for a desired output acceleration $\ddot{Y}$, as long as the matrix $H_v g$ may be inverted.

5.3.1.2 Stability Analysis

In this paragraph, the stability of the control system of the lighter than air robot is analyzed. The Lyapunov stability tools are used. There are basically two ways of using Lyapunov's direct method for control design and both are mostly trial and error methods. The first technique involves hypothesizing one form for the control law and then finding a Lyapunov function to justify the choice. The second technique in opposition, requires hypothesizing a Lyapunov function candidate and then finding a control law to make this candidate a real Lyapunov function. Perhaps, the most powerful approach to the study of stability of nonlinear systems is Lyapunov second method. It is a theoretical tool and the important feature is that it does not require the solutions of the system equations to be known, it requires generation of a Lyapunov function which can be very elusive.

The intuitive idea is that a physical system can only store finite energy and thus if it can be shown that energy is always being dissipated except at the equilibrium point, then the system must finally reach equilibrium when the energy is gone. The mathematical representation of the system's energy is in the Lyapunov function. In addition to stability tests, Lyapunov theory can be an aid to the designer for choosing control alternatives for nonlinear systems by finding a control law which yields a Lyapunov function for the system.

The resulting closed loop model is assumed linear and described by

$$\begin{aligned} \dot{X}_m &= AX_m + BY_{ref} \\ Y_m &= CX_m \end{aligned} \tag{5.58}$$

where A and B are constant matrices, $X_m = [V_m, P_m]$ is the state vector and $Y_m = P_m$ is the model output. The eigenvalues of the matrix A have negative real parts so that the model reference system has an asymptotically stable equilibrium state.

The desired acceleration $\ddot{Y}_m$ is then given by:

$$\ddot{Y}_m = CA^2X_m + CABY_{ref} + CB\dot{Y}_{ref} \tag{5.59}$$

Defining the augmented model output as:

$$\mathbf{Y}_m = \begin{pmatrix} \dot{Y}_m \\ Y_m \end{pmatrix} = DX_m + EY_{ref} \tag{5.60}$$

where

$$D = \begin{pmatrix} CA \\ C \end{pmatrix} \qquad E = \begin{pmatrix} CB \\ 0 \end{pmatrix} \tag{5.61}$$

We have:

$$\dot{Y}_m = D\dot{X}_m + E\dot{Y}_{ref} = A_Y Y_m + B_Y Y_{ref} + E\dot{Y}_{ref} \tag{5.62}$$

where

$$A_Y = DAD^{-1}$$
$$B_Y = \left(DB - DAD^{-1}E\right) \tag{5.63}$$

An augmented nonlinear output dynamics is obtained:

$$\dot{\mathbf{Y}}_m = \begin{pmatrix} \ddot{Y}_m \\ \dot{Y}_m \end{pmatrix} = F + GU \tag{5.64}$$

where

$$F = \begin{pmatrix} H_v f + H_p h \\ h \end{pmatrix} \qquad G = \begin{pmatrix} h_v g \\ 0 \end{pmatrix} \tag{5.65}$$

Let's now define the output error vector e by:

$$e = \begin{pmatrix} \dot{Y}_m \\ Y_m \end{pmatrix} - \begin{pmatrix} \dot{Y} \\ Y \end{pmatrix} \tag{5.66}$$

The differential equation for the error vector e is given by:

$$\dot{e} = \dot{Y}_m - \dot{Y} = A_Y e + A_Y Y - \dot{Y} + B_Y Y_{ref} + E\dot{Y}_{ref} \tag{5.67}$$

In the present problem, the error vector should be reduced to zero by a suitable control vector U. This controller should be such that at steady state $Y = Y_m$ and $\dot{Y} = \dot{Y}_m$ or $e = \dot{e} = 0$. Thus the origin $e = 0$ is an equilibrium state.

A convenient starting point in the synthesis of the control vector U is the construction of a candidate Lyapunov function for the system.

$$W_L(e) = e^T P e \tag{5.68}$$

where P is a real symmetric matrix. Taking the derivative of $W_L(e)$ with respect to time gives:

$$\dot{W}_L(e) = \dot{e}^T P e + e^T P \dot{e} = e^T \left(A_Y^T P + P A_Y\right) e + 2\hat{M} \tag{5.69}$$

where

$$\hat{M} = e^T P \left(A_Y Y - \dot{Y} + B_Y Y_{ref} + E\dot{Y}_{ref}\right)$$

The function $W_L(e)$ is a Lyapunov function if

1. $A_Y^T P + P A_Y = -Q$ is a negative definite matrix
2. the control U can be chosen to make the scalar quantity $\hat{M}$ non positive

Condition 1 can always be met by a proper choice of P, since the eigenvalues of A are assumed to have negative real parts, corresponding to the stability defined for the model. The problem is then to choose an appropriate control vector U so that $\hat{\mathrm{M}}$ is either zero or negative. The following case is for $\hat{M} = 0$

$$A_Y Y - \dot{Y} + B_Y Y_{ref} + E\dot{Y}_{ref} = 0$$

gives the following solution for U

$$U = G^{-1}\left[A_Y Y - F + B_Y Y_{ref} + E\dot{Y}_{ref}\right] \tag{5.70}$$

or equivalently:

$$U = \begin{pmatrix} (h_v g)^{-1} \\ 0 \end{pmatrix} \left(\dot{Y}_m - \begin{pmatrix} h_v f + h_p h \\ h \end{pmatrix} \right)$$
$$\text{when } \|U\| \le U_{max} \text{ unless } U = U_{max} \tag{5.71}$$

where $Y = Y_m$. This control law is exactly the one proposed in relation (5.71), which proves its asymptotic stability.

5.3.2 Trajectory Tracking in a High Constant Altitude Flight

Compared to satellites, the airship platform in the lower stratosphere has the advantages of being closer to the ground for better resolution images and requires less power for radio wave relay. Since a stratospheric LTA platform has to be light and large in its displacement volume, a non rigid structured hull would be most adequate. Winds in the stratosphere are weak at the altitude around 20 km above the ground, where the atmospheric pressure is about 40 hPa. The average temperature at this altitude is −50 degrees Celsius. Air density at this altitude is about 1/20 of that at sea-level and the LTA envelope needs to be large enough to yield necessary lift. This kind of airship is a super-light weight membrane structure. They are assumed super-pressurized, buoyant helium is expected to remain above atmospheric pressure inside the envelope, independent of variations in the environment of the airship. Neither the volume nor the mass changes during a flight, assuming that no ballast is dropped.

One of the basic problem is the control of this stratospheric airship. Some analysis were made in [48], dependent on assumptions made from linear models for each studied airship (moving in troposphere). The linear models were obtained from non linear simulation models by linearizing about a number of chosen trim speeds representative of a typical speed envelope. The decoupled linear models comprised the longitudinal and lateral motions of the neutrally buoyant airship, for speeds from the hover (0–0.1 m/s) to 10 m/s. In this paragraph, Input/Output linearization is proposed as a control method.

When the lighter than air robot flies at constant altitude, then the system dynamics can be expressed as [24]:

$$\begin{aligned} m_x\ddot{x} &= (F_1 + F_2)\cos\theta - k_x\dot{x} \\ m_y\ddot{y} &= (F_1 + F_2)\sin\theta - k_y\dot{y} \\ J_z\ddot{\theta} &= (F_1 - F_2)(k_t l_x + l_y) - k_\theta\theta \end{aligned} \tag{5.72}$$

where l_x and l_y are acting lengths and k_t is a rotor dependent constant. F_1 and F_2 are actuating forces of the motors. This airship can go forward or backward depending on the orientation of the actuators.

Neglecting the air resistance and all kind of damping and friction, and assuming than the added mass coefficients are identical in the x and y directions, the following relations can be proposed:

$$\begin{aligned}\ddot{x} &= u_1 \cos\theta \\ \ddot{y} &= u_1 \sin\theta \\ \ddot{\theta} &= u_2\end{aligned} \tag{5.73}$$

where

$$\begin{aligned}u_1 &= \frac{(F_1 + F_2)}{m} \\ u_2 &= (F_1 - F_2)\frac{k_t l_x + l_y}{J_z}\end{aligned} \tag{5.74}$$

This is an under actuated system. The non integrable condition arising in terms of acceleration is called the second order non holonomic condition.

5.3.2.1 Properties of the Dynamic Model

The aim of this paragraph is to study some properties of the model (5.73), such as controllability.

State-Space Formulation If the state-space variable is respectively defined as $X = (x\ y\ \theta\ \dot{x}\ \dot{y}\ \dot{\theta})^T$ and the input $U = (u_1\ u_2)^T$ then the state space model can be expressed as:

$$\begin{aligned}\dot{X} &= f_0(X) + g(X)U \\ Y &= h(X) = \begin{pmatrix} x \\ y \end{pmatrix}\end{aligned} \tag{5.75}$$

Where

$$f_0(X) = \begin{pmatrix} x_4 \\ x_5 \\ x_6 \\ 0 \\ 0 \\ 0 \end{pmatrix} \qquad g(X) = \begin{pmatrix} 0 & 0 \\ 0 & 0 \\ 0 & 0 \\ \cos(x_3) & 0 \\ \sin(x_3) & 0 \\ 0 & 1 \end{pmatrix} \tag{5.76}$$

A number of outputs must be chosen to be controlled equal to the number of its inputs to make the system be square, for this aim x, y are selected as outputs of interest.

These equations represent an affine system with drift. The vector fields f_0, g are smooth typically real analytic on R^6. The control U is a measurable function taking values in a compact subset $\mathbf{U} \subset R^2$ containing zero in its interior.

Drift less non-holonomic control systems have been studied by [39, 181]. Several important results have been derived based on the structure of Lie algebra generated by the control vector fields. The discussion of non-holonomic systems with drift has been concentrated on the dynamic extension of drift free systems.

Accessibility and Controllability Controllability indicates the existence of a path that connects an initial configuration to the desired final configuration, given a non holonomic system. There are many possible approaches to finding conditions for local controllability leading to different results and requiring different hypotheses. For analytic affine systems, the entire information about local properties of the system such as local controllability is contained in the values of the iterated Lie brackets of the vector fields f_0, g_1, g_2. Moreover, these values are easily computable. Therefore it is a natural approach to look for conditions for local controllability in terms of the elements of the Lie Algebra generated by the vector fields f_0, g_1, g_2.

In the sequel, the control characteristic indices σ_i, equal to the least order of the time derivative of the output Y which is directly affected by some input, are introduced.

$$Y_i^{(\sigma_i)} = L_{f_0}^{(\sigma_i)} + \sum_{j=1}^{m} L_{g_j} h_j u_j \qquad 1 \le i \le m = 2 \tag{5.77}$$

Proposition 5.4 *A system represented by the Eq.* (5.75) *is locally accessible and the non-holonomy order is* $r = 4$, *while the growth vector is* (2, 4, 6). *The relative growth vector is* (2, 2, 2). *The Control characteristic indices associated with system* (5.73) *are given as follows*:

$$\sigma_1 = \sigma_2 = 2 \tag{5.78}$$

Proof For this particular class of affine non-holonomic system, to check the controllability, the following Lie brackets must be considered.

$$\begin{aligned} g_3 &= [f_0, g_1] \\ g_4 &= [f_0, g_2] \end{aligned} \tag{5.79}$$

Where

$$ad_g^i g = [g_0, ad_g^{i-1} g] \tag{5.80}$$

with:

$$\begin{aligned} \mathrm{rank}(g_1, g_2) &= 2 \\ \mathrm{rank}(g_1, g_2, g_3, g_4) &= 4 \\ \mathrm{rank}(g_1, g_2, g_3, g_4, g_5, g_7) &= 6 \end{aligned} \tag{5.81}$$

Thus the nonholonomy order is 4 while the growth vector is $(2, 4, 0, 6)$ and the relative growth vector is $(2, 2, 0, 2)$. It has been checked the Lie algebra spans the entire space ($\Re^6$). Thus the system is locally accessible.

$$\begin{aligned} L_{g_j} L_g^k h_i(X) = 0 \quad \texttt{with } 1 \le j \le m \texttt{ and } 0 \le k \le \rho_i - 2 \\ L_{g_j} L_g^{\rho_i - 1} h_i(X) \neq 0 \quad \texttt{for some } j \texttt{ such that } 1 \le j \le m \end{aligned} \tag{5.82}$$

where

$$\begin{aligned} L_f h &= \frac{\partial h}{\partial x} f \\ L_f^0 h &= h \\ L_f^i h &= L_f \left(L_f^{i-1} h \right) \end{aligned} \tag{5.83}$$

The above definition of control characteristic indices is given about the origin: it may be given around any point $\bar{X} \in R^n$ such that $\text{rank}(G(\bar{X})) = 2$ and $\text{rank}(dh_1(\bar{X}), dh_2(\bar{X}), \ldots, dh_m(\bar{X})) = 2$

$$\begin{pmatrix} \ddot{x} \\ \ddot{y} \end{pmatrix} = \Phi_1(X) \begin{pmatrix} \Gamma_1 \\ \Gamma_2 \end{pmatrix} \tag{5.84}$$

with

$$\Phi_1(X) = \begin{pmatrix} a_1 & a_1 \\ -a_2 & a_2 \end{pmatrix}$$

Thus the relative degree is:

$$\sigma_1 = \sigma_2 = 2$$

□

As $\sigma = \sum_{i=1}^{2} \sigma_i = 4$, there exists no zero dynamics. In this case, the controllability indices are equal to the control characteristic indices. The system is of minimum phase.

5.3.2.2 Motion Control

The next step is to control the motion of the vehicle onto the path. For kinematics models, the stabilization problem has essentially been solved with two types of control laws:

- time-varying piecewise continuous control.
- Time-varying continuous control.

An analogous study must be made for dynamics models. Given a desired state trajectory, a controller which stabilizes the system to this trajectory, must be constructed. The system given by relation (5.75) is not input—state linearizable. However, this system having a well defined relative degree can be Input/Output linearizable. The key of this method is to transform the non linear system into a linear one by applying a state feedback and state transformation.

Proposition 5.5 *If the control U is chosen such that*

$$U = E(X)^{-1}(-a(X) + V) \tag{5.85}$$

Then the system (5.73) *can be equivalently written as*

$$\ddot{y} = V \tag{5.86}$$

Where V is the new input.

If the decoupling matrix $E(X)$ is non singular then the system is locally decouplable and Input Output linearizable by state feedback

$$\begin{aligned} E(X) &= \frac{\partial h}{\partial X}\frac{\partial f_0}{\partial X} f_0 \\ a(X) &= \frac{\partial h}{\partial X}\frac{\partial f_0}{\partial X} g \end{aligned} \tag{5.87}$$

The output to be controlled is the output of a chain of cascaded integrators fed by a nonlinear but invertible forcing term.

For the airship,

$$r_1 = r_2 = 2$$

and

$$E(X) = \begin{pmatrix} \cos(x_3) & 0 \\ \sin(x_3) & 0 \end{pmatrix} \tag{5.88}$$

where E is a singular matrix. To skip this problem, differentiating the outputs four times allow all inputs to appear but in this case the determinant of the decoupling matrix depend on the force u_1, this causes a serious problem when $u_1 = 0$. This drawback can be avoided by choosing the point of interest different from the gravitational center of airship. Let:

$$\begin{aligned} x_p &= x + l\cos(x_3) \\ y_p &= y + l\sin(x_3) \end{aligned} \tag{5.89}$$

be the coordinate of this point X_p.

In this case the drift term can be written as:

$$f(X) = \begin{pmatrix} x_4 \\ x_5 \\ x_6 \\ x_6^2 l\cos(x_3) \\ x_6^2 l\sin(x_3) \\ 0 \end{pmatrix} \tag{5.90}$$

$$h(X) = \begin{pmatrix} x_1 \\ x_2 \end{pmatrix} \tag{5.91}$$

applying the same calculus, $r_1 = r_2 = 2$ with

$$\begin{aligned} E(X) &= \begin{pmatrix} \cos(x_3) & -l\sin(x_3) \\ \sin(x_3) & l\cos(x_3) \end{pmatrix} \\ a(X) &= \begin{pmatrix} -x_6^2 l\cos(x_3) \\ -x_6^2 l\sin(x_3) \end{pmatrix} \end{aligned} \tag{5.92}$$

and

$$\det(E(X)) = l$$

The linearized model system does not contain an unobservable zero dynamics. Thus, using a stable tracking law, we can make the point X_p tracking the reference trajectory.

Proof Input-output linearization of the system is obtained by differentiating the outputs y_i until the inputs appear. So, the system become:

$$y_i^{r_i} = L_f^{r_i} h_i + \sum_{j=1}^{2} L_{g_i} L_g^{r_i} h_j u_j \tag{5.93}$$

with

$$L_{g_i} L_g^{r_i} h_j \neq 0$$

for at least one j. Rewrite those equations in the compact form

$$\begin{pmatrix} y_1^{r_1} \\ y_2^{r_2} \end{pmatrix} = \begin{pmatrix} L_f^{r_1} h_1 \\ L_f^{r_2} h_2 \end{pmatrix} + \begin{pmatrix} L_{g_1} L_f^{r_1} h_1 & L_{g_2} L_f^{r_2} h_2 \\ L_{g_1} L_f^{r_1} h_1 & L_{g_2} L_f^{r_2} h_2 \end{pmatrix} \begin{pmatrix} u_1 \\ u_2 \end{pmatrix} \tag{5.94}$$

and the control canceling the nonlinearity is given by the equation:

$$U = E(X)^{-1} \begin{pmatrix} \nu_1 - L_f^{r_1} \\ \nu_2 - L_f^{r_2} \end{pmatrix} \tag{5.95}$$

where E(x), the decoupling matrix, must be non-singular, ν_i are the auxiliary inputs can be chosen as follows:

$$\nu = yd_i^{r_i} - k_1^i e_i^{r_i - 1} - \ldots - k_{r_i}^i e_i \tag{5.96}$$

with

$$e_i = yd_i - y_i$$

yd being the reference trajectory. The coefficients k_i^j must be chosen such that the polynoms:

$$e_i^{r_i} + k_1^i e_i^{r_i - 1} + \ldots + k_{r_i}^i e_i = 0$$

is Hurwitz to guarantee the asymptotic convergence of the errors to zero. □

5.3.3 Variable Structure Robust Control

In a variety of missions, inertial trajectory control of lighter than air robot in three-dimensional space is essential. The control model can be written as:

$$\begin{aligned}
\dot{x} &= V \sin\chi \cos\gamma \\
\dot{y} &= V \cos\chi \cos\gamma \\
\dot{z} &= V \sin\gamma \\
\dot{V} &= \frac{(T\cos\alpha - D + (B - mg)\sin\gamma)}{m + m_{11}} \\
\dot{\chi} &= \frac{(L + T\sin\alpha)\sin\sigma}{(m + m_{22})V\cos\gamma} \\
\dot{\gamma} &= \frac{-(L\cos\sigma + T\cos\sigma\sin\alpha + (B - mg)\cos\gamma)}{(m + m_{33})V}
\end{aligned} \tag{5.97}$$

The design of flight path controllers is relatively complicated. The reason lies in the higher relative degrees (4, 4, 4) (ignoring small forces contributed by the control surfaces) of the position coordinates (x, y, z) with second derivative of thrust and control surface deflections as control inputs. Each of the variables χ, γ has relative degree only 2. Relative degree of an output of a system is defined to be the order of the derivative of the output in which control input appears for the first time. Thus, dynamic inversion approach requires derivative of fourth order of x, y, and z. The presence of uncertainties in the model creates additional difficulties.

Remark 5.5 One can use an adaptive back stepping design for the compensation of uncertainties. However, computational difficulties arise because the virtual control inputs appear and depend on uncertain aerodynamic parameters.

The approach followed in this section is robust control design using the Variable Structure Control (VSC) theory, an appropriate sliding surface on which the trajectory has desirable property, is first selected.

The control input is taken as $U = (\dot{T}, \dot{\sigma}, \dot{\alpha})^T$, the output $Y = (x, y, z)^T$ and the state vector $X = (X_1, X_2)^T$, $X_1 = (x, y, z)^T$, $X_2 = (V, \chi, \gamma)^T$.

The following sliding surface is chosen:

$$S = \ddot{e} + K_v \dot{e} + K_p e + K_i \int e dt \tag{5.98}$$

where $e = (x - x_r, y - y_r, z - z_r)^T = X_1 - X_{1r}$.

These gains K_p, K_v, K_i are chosen so that $S = 0$ yields exponentially stable response for e; Integral feedback provides additional flexibility for a robust design. The motion of the closed loop system including the VSC law evolves in two phases.

- First, the trajectory beginning from arbitrary initial state is attracted toward S=0.
- In the second phase, which is termed a sliding phase, e converges to zero because $S = 0$.

Once the choice of a sliding surface has been made, a controller must be designed such that $S = 0$ becomes an attractive surface; Differentiating X twice gives:

$$\ddot{X} = B_1 \begin{pmatrix} \dot{V} \\ \dot{\chi} \\ \dot{\gamma} \end{pmatrix} = B_1 \left(f_0 \left(V, \gamma \right) + f_1 \left(T, V, \gamma \right) \right) \tag{5.99}$$

f_0, f_1 are non linear functions given respectively by:

$$f_0(X_2) = \begin{pmatrix} \frac{(B-mg)\sin\gamma}{m+m_{11}} \\ 0 \\ -\frac{(B-mg)\cos\gamma}{(m+m_{33})V} \end{pmatrix}$$

$$f_1(X_4) = \begin{pmatrix} \frac{(T\cos\alpha - D)}{m+m_{11}} \\ \frac{(T\sin\alpha + L)\sin\sigma}{(m+m_{22})V\cos\gamma} \\ \frac{((T\sin\alpha + L)\cos\sigma)}{(m+m_{33})V} \end{pmatrix}$$

and the matrix B_1 can be written as:

$$B_1 = \begin{pmatrix} \sin\chi\cos\gamma & V\cos\chi\cos\gamma & -V\sin\chi\sin\gamma \\ \cos\chi\cos\gamma & V\sin\chi\sin\gamma & -V\cos\chi\sin\gamma \\ \sin\gamma & 0 & V\cos\gamma \end{pmatrix} \tag{5.100}$$

Differentiating one more time, the following relation is obtained

$$X^{(3)} = \dot{B}_1(f_0 + f_1) + B_1\left(\dot{f}_1 + \dot{f}_2\right) = f_3 + B_1 B_2 U \tag{5.101}$$

where

$$f_3 = \dot{B}_1(f_0 + f_1) + B_1\dot{f}_0 + B_1\left(\frac{\partial f_1}{\partial V}\dot{V} + \frac{\partial f_1}{\partial \chi}\dot{\chi} + \frac{\partial f_1}{\partial \gamma}\dot{\gamma}\right) \tag{5.102}$$

$$B_4 = B_1 B_2 \tag{5.103}$$

with

$$B_2 = \begin{pmatrix} \frac{1}{m+m_{11}} & 0 & \frac{-1}{m+m_{11}}\frac{\partial L}{\partial \alpha} \\ 0 & \frac{L}{m+m_{22}}\frac{\cos\sigma}{V} & 0 \\ 0 & \frac{mL}{V(m+m_{33})}\frac{\sin\sigma}{\cos\gamma} & \frac{m\cos\sigma}{(m+m_{33})V\cos\gamma}\frac{\partial L}{\partial \alpha} \end{pmatrix} \tag{5.104}$$

Differentiating S given in relation (5.98) and using relation (5.101) gives

$$\begin{aligned}\dot{S} &= f_3 + B_4 U + K_v B_1 (f_0 + f_1) + K_p \dot{e} - K_i e - X_r^{(3)} - K_v \ddot{X}_r \\ &= f_5 + B_4 U = \left(f_5^* + \Delta f_5\right) + \left(B_4^* + \Delta B_4\right) U \end{aligned} \tag{5.105}$$

Here starred functions denote nominal values of functions and Δf_5, ΔB_4, denote uncertain functions.

For the existence of a VSC (or even a feedback linearizing control), invertibility of matrix B_4 is required. Of course, B_4 is function of its arguments and singularities of this matrix belong to only certain hypersurfaces. In view of relation (5.104), if B is singular, U cannot directly affect S, computing the determinant gives:

$$\det(B_4) = -\frac{\partial L}{\partial \alpha} \frac{L \cos^2 \alpha}{m} \tag{5.106}$$

This determinant is different from zero if $\sigma \neq \pm\pi/2$ rad.

Remark 5.6 Thus in a neighborhood of the trim value, B is invertible and a variable structure control law can be designed. Of course computation of exact region in which it is invertible is interesting, but it is quite involved because B_4 is a complicated nonlinear function. For trajectory control, maneuver through the region in which singularity lies must be avoided by a proper trajectory planner.

For the derivation of the control law, the Lyapunov approach is used. If there exists a scalar function $W(x, t)$ having continuous first partial derivatives and satisfying the conditions

- $W(x, t)$ is positive definite
- $\dot{W}(x, t)$ is negative definite

thus the equilibrium state at the origin is uniformly asymptotically stable and W is called a Lyapunov function.

For the lighter than air robot, the following candidate Lyapunov function is proposed:

$$W = \frac{1}{2} S^T S \tag{5.107}$$

The derivative of W is given by:

$$\dot{W} = S^T \left(f_5 + \Delta f_5 + (B_4 + \Delta B_4) U\right) \tag{5.108}$$

In view of relation (5.108) for making $\dot{W}$ negative, the control law is chosen of the form

$$U = (B_4)^{-1} \left(-f_5 - K S - K' \operatorname{sign}(S)\right) \tag{5.109}$$

Where K is a diagonal matrix, and $K > 0$ is yet to be chosen. Now the following assumption is made

$$\begin{aligned} \left\| \Delta B_4 (B_4)^{-1} \right\|_\infty &\le a_0 \\ \left\| \Delta f_5 - \Delta B_4 (B_4)^{-1} (f_5 + KS) \right\|_\infty &\le a_0 \end{aligned} \tag{5.110}$$

The parameters a_0 and a_1 can be taken as positive constants. The first inequality restricts the uncertainty in the input matrix which requires that any uncertain U dependent term appearing in relation (5.109) should be small. A good estimate of this term can be used to satisfy assumption (5.110). In view of relation (5.109) for compensating the uncertainty, one chooses the gain K as

$$K \ge (1 - a_0)^{-1} (\eta + a_1) \eta > 0 \tag{5.111}$$

Then using the uncertainty bounds of assumption 1, one can show that

$$\dot{W} \ge -\eta \left\| S \right\|_1 - S^T K S \tag{5.112}$$

$\forall S \ne 0$ almost everywhere on $t \in [0, +\infty)$ where $\|S\|_1 = \sum_1^3 |S_i|$ Thus the surface $S = 0$ is reached in a finite time and by the definition of S, the trajectory tracking error e tends to zero. It is noted that the VSC law, can be synthesized using the computed value of $\ddot{S}$ using the nominal values of the parameters. As the control system includes discontinuous functions, this can cause chattering phenomenon, here it can be avoided by smoothing the discontinuous functions by replacing sign function by saturation function. This modifies the control law in small boundary layers surrounding $S = 0$. Outside the boundary layers, relation (5.109) is valid and as ε tends to zero, the control laws with the sign functions are recovered.

5.3.4 Back Stepping Controller Design

The separation of the lighter than air robot model in dynamic and kinematic equations in a cascaded system appearance suggests that the back stepping solution which is a Lyapunov based control design approach can be applied. By formulating a scalar positive function of the system states and then choosing a control law that makes this function decrease, the nonlinear control system thus designed will be asymptotically stable. Moreover, it will be robust to some unmatched uncertainties.

Back stepping approach is a recursive procedure which allows deriving control law for a nonlinear system, based on an appropriate Lyapunov function candidate. Three approaches are presented below: the first one considers longitudinal/lateral decoupling, the second one uses the 6 degrees of freedom dynamics with Euler angles kinematic representation while the third utilizes quaternion kinematic representation.

5.3.4.1 Longitudinal/Lateral Decoupling

If the state dynamics involved in longitudinal and lateral motions appear to be weakly dependent, decoupling between lateral and longitudinal states is performed.

To describe the dynamics with the longitudinal plane, the following state variable $\eta_{long} = (x, z, \theta)$ and $\nu_{long} = (u, w, q)$ are used.

To describe the dynamics with the lateral plane, the following state variable $\eta_{lat} = (y, \phi, \psi)$ and $\nu_{lat} = (v, p, r)$ are used.

Longitudinal Control Four phases are to be considered within the longitudinal flight: take-off, transient longitudinal flight, steady longitudinal flight and landing. The take-off can be defined as a trajectory tracking and may be viewed as a part of the transient longitudinal phase. The control objective of the transient phase is to drive the lighter than air robot up to a reference altitude, cruise velocity and constant pitch angle θ_e. During this phase, the robot remains sufficiently close to its steady state to perform a longitudinal trajectory tracking. The control objective of landing is to reduce progressively the altitude and velocity until the lighter than air robot stops.

The longitudinal kino-dynamics model can be expressed as

$$
\begin{aligned}
\dot{x} &= u\cos\theta + w\sin\theta \\
\dot{z} &= -u\sin\theta + w\cos\theta \\
\dot{\theta} &= q \\
m_x\dot{u} + ma_x\dot{q} &= -m_x wq + \tfrac{1}{2}\rho C_D V^2 vol^{2/3} \\
&\quad - (mg - F_B)\sin\theta + F_M\cos\mu \\
m_z\dot{w} - ma_x\dot{q} &= -m_z uq + \tfrac{1}{2}\rho C_L V^2 vol^{2/3} + (mg - F_B)\cos\theta \\
&\quad - F_M\sin\mu + f_{el}\delta_{el} \\
J_y\dot{q} + ma_x\dot{u} - ma_z\dot{w} &= -ma_x uq - ma_z wq + \tfrac{1}{2}\rho C_m V^2 vol \\
&\quad - mga_z\sin\theta - a_x F_B\cos\theta \\
&\quad - F_M O_x\sin\mu + F_M O_z\cos\mu + M_{el}\delta_{el}
\end{aligned}
\tag{5.113}
$$

V is the relative lighter than air robot velocity, *vol* is the volume of the hull, (a_x, a_z) and (O_x, O_z) determine respectively the position of the center of mass and the vectored thrust with respect to the body fixed frame $\mathcal{R}_m$, $f_{el}\delta_{el}$ and $M_{el}\delta_{el}$ are the force and moment induced by the elevator, δ_{el} is the elevator angle.

The proposed trajectory tracking strategy is based on backstepping techniques. Let η_{ref} represent the expected lighter than air robot position and $e_1 = \eta - \eta_{ref}$ the position error. To perform the tracking, an additional variable e_2 that will behave as a sliding variable is introduced:

$$e_2 = \dot{e}_1 + \lambda e_1$$

then $e_2 = \dot{\eta} - \dot{\eta}_{ref} + \lambda e_1 = \dot{\eta} - \dot{\eta}_v$ where $\dot{\eta}_v = \dot{\eta}_{ref} - \lambda e_1$.

In the sequel e_2 is considered as a velocity error and $\dot{\eta}_v$ as a virtual velocity. From the relation $\dot{\eta}_v = J(\eta)\nu_v$ and $\dot{\eta}_{ref} = J(\eta)\nu_{ref}$, the virtual velocity can be expressed with respect to the body fixed frame $\mathcal{R}_m$.

$$\nu_v = \nu_{ref} - J^{-1}(\eta)\lambda e_1$$

Following the backstepping scheme, which consists of two embedded control loops, ν_v can be viewed as a kinematic controller. Considering the positive functions $S_1 = \frac{1}{2}\|e_1\|^2$ whose time derivative verifies $\dot{S}_1 = e_1^T \dot{e}_1 = -e_1^T \lambda e_1 + e_1^T e_2$. Once $\dot{\eta} = \dot{\eta}_v$ or $\nu = \nu_v$, $e_2 = 0$, the position error e_1 converges asymptotically to zero.

The next step is then to determine the dynamic controller which allows to stabilize e_2 to zero.

Introducing $e'_1 = J^{-1}(\eta)e_1$ and $e'_2 = J^{-1}(\eta)e_2 = \nu - \nu_v$, the controller is defined by:

$$T_p = T_d(\nu) + T_a(\nu) + T_s(\eta) + \mathbf{M}\dot{\nu}_v - \lambda_2 e'_2 \quad \lambda_2 > 0$$

The closed loop stability is deduced from the following candidate Lyapunov function:

$$S_2 = \frac{1}{2} e_2'^T \mathbf{M} e'_2$$

The parameter $\mathbf{M}$ being the inertia matrix.

The time derivative can be expressed as

$$\dot{S}_2 = e_2'^T \mathbf{M}\dot{e}_{2'} = e_2'^T \left[T_p - T_d(\nu) - T_a(\nu) - T_s(\eta) - \mathbf{M}\dot{\nu}_v\right]$$

Replacing T_p by its closed loop expression, $\dot{S}_2 = -e_2'^T \lambda_2 e_{2'} < 0$. Showing that $e_{2'}$ converges asymptotically to zero. Introducing $a = \dot{\nu}_v - \mathbf{M}^{-1}\lambda_2 e_2$, the transition input between the kinematic and the dynamic loop, the controller expresses

$$T_p = T_d(\nu) + T_a(\nu) + T_s(\eta) + \mathbf{M}a$$

On the other hand, T_p is directly related to the inputs: the elevator angle and the horizontal and vertical components of the thrust by the relation.

$$T_p = BU$$

where

$$B = \begin{pmatrix} 1 & 0 & 0 \\ 0 & -1 & f_{el} \\ O_z & -O_x & M_{el} \end{pmatrix}, \qquad U = \begin{pmatrix} F_{x_{vect}} \\ F_{z_{vect}} \\ \delta_{el} \end{pmatrix}$$

When the lighter than air velocity is different from zero, the matrix B is invertible and the system is controllable. This condition is naturally verified as soon as the tracking phase is started.

Lateral Control This flight phase is used to control the lighter than air robot lateral motion after a longitudinal flight. The altitude, pitch angle and velocity are initially stabilized to their reference equilibrium values.

The lateral kinodynamics can then be expressed as

$$
\begin{aligned}
\dot{y} &= u_e \cos\theta_e \cos\psi + v \sin\psi \\
\dot{\phi} &= p \\
\dot{\psi} &= r \\
m_y \dot{v} - m a_x \dot{p} + m a_x \dot{r} &= m_x w_e p - m_x u_e r + F_{Y_e} + F_{Y_v} v + F_{Y_p} p + F_{Y_r} r \\
&\quad + (mg - B)\phi \cos\theta_e + f_{Y_{rud}} \delta_{rud} \\
J_x \dot{p} - J_{xz} \dot{r} - m a_x \dot{v} &= -m a_x w_e p + m a_x u_e r + M_{L_e} + M_{L_v} v \\
&\quad + M_{L_p} p + M_{L_r} r - a_x m g \phi \cos\theta_e \\
J_z \dot{r} - J_{xz} \dot{p} + m a_x \dot{v} &= m a_z w_e p - m a_x u_e r + M_{N_e} + M_{N_v} v + M_{N_p} p \\
&\quad + M_{N_r} r + a_x m g \phi \cos\theta_e + M_{N_{rud}} \delta_{rud}
\end{aligned}
\tag{5.114}
$$

where J_x, J_y, J_{xz} are inertia parameters, $F_{Y_e} + F_{Y_v} v + \dots$ are linearized expressions of aerodynamic forces and moments, $f_{Y_{rud}} \delta_{rud}, M_{N_{rud}} \delta_{rud}$ are the aerodynamic force and moment induced by the mobile surface, δ_{rud} is the deflection angle of the rudder.

Remark 5.7 For this model, the aerodynamic stability condition is expressed by $F_{Y_e} = M_{L_e} = M_{N_e} = 0$.

To consider separately the kinematic and the dynamic regulation, the lateral motion is based on a backstepping control scheme, through a path following scheme.

Let L be the lateral distance between the bow of the lighter than air robot and the desired path and $\tilde{\psi}$ be the angular error between the velocity V and the mobile frame X_d axis whose orientation is given by ψ_d. The error dynamics reduces to

$$
\begin{aligned}
\dot{L} &= V \sin\bar{\psi} \\
\dot{\tilde{\psi}} &= r_c
\end{aligned}
\tag{5.115}
$$

where $\bar{\psi} = \psi + \beta - \psi_d$, β being the skid angle and r_c the control variable. To stabilize the system, the following controller is proposed

$$
r_c = -K_1 L V \frac{\sin\bar{\psi}}{\bar{\psi}} - K_2 V \bar{\psi}
$$

The stability of the closed loop system is investigated through the following candidate Lyapunov function.

$$
S_1 = \frac{1}{2} K_1 L^2 + \frac{1}{2} \bar{\psi}^2 > 0
$$

By differentiation versus time,

$$\dot{S}_1 = K_1 L\dot{L} + \bar{\psi}\dot{\bar{\psi}} = K_1 L V \sin\bar{\psi} + r_c\bar{\psi} = -K_2 V\dot{\psi}^2 \leq 0$$

Thus $\bar{\psi}$ converges asymptotically to zero. By using some additional boundedness arguments and Barbalat's lemma, the convergence of L to zero can be proven as well.

Let $\tilde{r} = r - r_c$ denote the angular velocity error. Considering the dynamic model with the kinematic equation $\dot{\psi} = p$, the lateral dynamic can be expressed as:

$$\begin{aligned} M^* X &= AX + B\delta_{rud} \\ r &= CX \end{aligned} \tag{5.116}$$

where the following notations are used: $X = (v, p, r, \phi)^T$ and $C = (0, 0, 1, 0)$.

The following controller is considered

$$\delta_{rud} = \left(C M^{*^{-1}} B\right)^{-1}\left(-C M^{*^{-1}} AX + \dot{r}_c - K_3\bar{r}\right)$$

Differentiating the output equation,

$$\dot{r} = C\dot{X} = C M^{*^{-1}} AX + C M^{*^{-1}} B\delta_{rud}$$

in which $C M^{*^{-1}} B$ is a scalar. Then,

$$\dot{S}_2 = \bar{r}\left(\dot{r} - \dot{r}_c\right) = \bar{r}\left(C M^{*^{-1}} AX + C M^{*^{-1}} B\delta_{rud} - \dot{r}_c\right) = -K_3\bar{r}^2 \leq 0$$

This proves that the angular velocity is asymptotically stabilized.

5.3.4.2 Euler Angles Approach

A nonlinear model based velocity and attitude controller is designed by using the backstepping technique. The equations of motion are defined as

$$\mathbf{M}\dot{\nu} + C(\nu)\nu + D(\nu)\nu + \tau_s(\eta_2) = \tau_p + \mathbf{R}^{-1}b \tag{5.117}$$

where the notations are reminded: $\nu = (u, v, w, p, q, r)^T$ with $\eta = (\eta_1, \eta_2)^T$ where $\eta_1 = (x, y, z)^T$ and $\eta_2 = (\phi, \theta, \psi)^T$.

A non linear model based velocity and attitude controller is designed by using the backstepping technique. The output to be controlled is redefined from position and attitude to linear velocity and attitude. By feeding the controller the appropriate guidance signals, positional convergence is ensured such that the guidance based path following task objectives are met.

Start by defining the projection matrix

$$H = \begin{bmatrix} 0_{3\times 3} & I_{3\times 3} \end{bmatrix}$$

Then the error variables are defined as

$$z_1 = \eta_2 - \eta_{ref} \qquad z_2 = \nu - \nu_{ref} \tag{5.118}$$

where $\nu_d \in \Re^6$ is a vector of stabilizing functions to be defined later.

The following control Lyapunov function is proposed

$$V_1 = \frac{1}{2} z_1^T K_1 z_1 \tag{5.119}$$

where $K_1 = k_1^T > 0$. Differentiating V_1 with respect to t along the z_1 dynamics yields

$$\begin{aligned} \dot{V}_1 &= z_1^T K_1 \dot{z}_1 = z_1^T K_1 \left(\dot{\eta}_2 - \dot{\eta}_{ref}\right) = z_1^T K_1 \left(J \nu_2 - \dot{\eta}_{ref}\right) \\ &= z_1^T K_1 J H z_2 + z_1^T K_1 \left(J \alpha_\omega - \dot{\eta}_{ref}\right) \end{aligned} \tag{5.120}$$

where $\alpha_\omega = H \nu_{ref}$. This motivates the choice of α_ω as

$$\alpha_\omega = J^{-1} \left(\nu_{ref} - z_1\right)$$

which results in

$$\dot{V}_1 = -z_1^T K_1 z_1 + z_2^T H^T J^T K_1 z_1 \tag{5.121}$$

V_1 is augmented to obtain

$$V_2 = V_1 + \frac{1}{2} z_2^T \mathbf{M} z_2 + \frac{1}{2} \underline{b}^T \Gamma^{-1} \underline{b}$$

where $\underline{b} \in \Re^6$ is an adaptation error defined as $\underline{b} = \hat{b} - b$ with $\hat{b}$ being the estimate of b and by assumption $\dot{b} = 0$. The symmetric definite positive matrix Γ is the adaptation gain matrix.

Differentiating V_2 along the trajectory z_1, z_2 and $\underline{b}$, the following relation is obtained:

$$\dot{V}_2 = -z_1^T K z_1 + z_2^T \left(H^T T^T K_1 z_1 + M \dot{z}_2\right) + \underline{b}^T \Gamma^{-1} \dot{\hat{b}}$$

Since $\mathbf{M} = \mathbf{M}^{-1}$ and $\dot{\hat{b}} = \dot{\underline{b}}$,

$$\mathbf{M} \dot{z}_2 = \mathbf{M} \left(\nu - \nu_{ref}\right) = \tau + J^{-1} b - g - D\nu - C\nu - \mathbf{M} \dot{\nu}_{ref}$$

which yields to

$$\begin{aligned} \dot{V}_2 = &-z_1^T K_1 z_1 + z_2^T \left(H^T T^T K_1 z_1 + \tau + J^{-1} b - g\right) \\ &+ z_2^T \left(-D\nu - C\nu - M \dot{\nu}_{ref}\right) + \underline{b}^T \Gamma^{-1} \dot{\hat{b}} \end{aligned}$$

Using the fact that $V = z_2 + \nu_{ref}$ and $b = \hat{b} - \underline{b}$,

$$\dot{V}_2 = -z_1^T K_1 z_1 - z_2^T D z_2 - z_2^T C z_2 + z_2^T \left(H^T T^T K_1 z_1 + \tau + J^{-1}\hat{b} - g\right)$$
$$+ z_2^T \left(-D\nu_{ref} - C\nu_{ref} - \mathbf{M}\dot{\nu}_{ref} + \underline{b}^T \Gamma^{-1}\left(\dot{\hat{b}} - \Gamma J^{-T} z_2\right)\right)$$

where $z_2^T C z_2 = 0$, since C is skew symmetric. By assigning,

$$\tau = \mathbf{M}\dot{\nu}_{ref} + C\nu_{ref} + D\nu_{ref} + g - J^{-1}\hat{b} - H^T T^T K_1 z_1 - K_2 z_2$$

where $K_2 = K_2^T > 0$ and by choosing $\dot{\hat{b}} = \Gamma J^{-T} z_2$, the following relation is obtained:

$$\dot{V}_2 = -z_1^T K_1 z_1 - z_2^T (D + K_2) z_2$$

Thus the error asymptotically converges to zero.

5.3.4.3 Quaternion Approach

The vectorial back stepping gives an advantage over the integrator back stepping in case of applying to the complicated MIMO system since the control law is derived in few steps [7–9].

In a low airspeed or hovering state, the aerodynamics are reduced, and consequently the control surfaces authority, thus demanding a superior action by the motors for position and attitude control. This might lead to saturation of the control signals, which are usually bounded. The key issue is then to include these input limitations in the controller synthesis. The LTAR dynamics can be presented in the following form:

$$\mathbf{M}\dot{X} = -\Omega_6 \mathbf{M} X + E S a_g + f \tag{5.122}$$

where $E = \binom{m_w I_3}{m c_3}$

$$\Omega_6 = \begin{pmatrix} \Omega_3 & 0 \\ 0 & \Omega_3 \end{pmatrix}$$

with $\Omega_3 = Sk(\omega)$, $c_3 = Sk(c)$

$$\Omega_7 = \begin{pmatrix} \Omega_3 & 0 \\ 0 & \frac{1}{2}\Omega_4 \end{pmatrix}$$

with

$$\Omega_4 = \begin{pmatrix} 0 & -\Omega_1 & -\Omega_2 & -\Omega_3 \\ \Omega_1 & 0 & -\Omega_3 & \Omega_2 \\ \Omega_2 & \Omega_3 & 0 & -\Omega_1 \\ \Omega_3 & -\Omega_2 & \Omega_1 & 0 \end{pmatrix}$$

$$\begin{aligned}\dot{X} &= KX + \mathbf{M}^{-1}\left(ESa_g + f\right)\\ \dot{\eta} &= DCX\\ \dot{S} &= -\Omega_3 S\\ \dot{D} &= D\Omega_7\end{aligned} \tag{5.123}$$

where $K = -M^{-1}\Omega_6 M$ linearly depends on the velocity V, whereas M is constant or slowly varying.

Consider two intermediate output variables.

$$\begin{aligned}Y_1 &= a\eta + TX\\ Y_2 &= \Xi X\end{aligned} \tag{5.124}$$

where a is a positive scalar and the candidate Lyapunov function is defined as:

$$W = \frac{1}{2}Y_1^T Y_1 + \frac{1}{2}Y_2^T Y_2 \tag{5.125}$$

Then

$$\dot{W} = Y_1^T \dot{Y}_1 + Y_2^T \dot{Y}_2 \tag{5.126}$$

$$\begin{aligned}\dot{Y}_1 &= a\dot{\eta} + \dot{T}X + T\dot{X} = aTX + D\Omega_7 CX + T\dot{X}\\ \dot{Y}_2 &= \Xi \dot{X}\end{aligned} \tag{5.127}$$

Leading to

$$Y_2^T \dot{Y}_2 = (\Xi X)\,\Xi \dot{X} = X^T \Xi^2 \dot{X} = X^T T^T \dot{X}$$

or

$$\begin{aligned}\dot{W} = {} & (a\eta + 2TX)^T \left(aTX + D\Omega_7 CX + T\dot{X}\right)\\ & - (TX)^T (aTX + D\Omega_7 CX)\end{aligned} \tag{5.128}$$

If the input is chosen so that

$$T\dot{X} = (a\eta + 2TX) - aTX - D\Omega_7 CX \tag{5.129}$$

then

$$\dot{W} = -(a\eta + 2TX)^T (a\eta + 2TX) - a(TX)^T (TX) - X^T C^T \Delta_7^2 \Omega_7 CX \tag{5.130}$$

Taking into account that $\Delta_7^2 \Omega_6$ is anti-symmetric, which eliminates the last term, the derivative may finally be expressed as:

$$\dot{W} = -(a\eta + 2TX)^T (a\eta + 2TX) - a(TX)^T (TX) \tag{5.131}$$

which states that if $a > 0$ this derivative is negative definite, indicating a closed loop system with global asymptotic stability.

The closed loop dynamics are then given by the control law definition

$$\begin{aligned} T\dot{X} &= -(a+2)TX - a\eta - D\Omega_7 CX \\ \dot{\eta} &= TX \end{aligned} \tag{5.132}$$

or introducing $\Lambda_1^2 = -(a+2)I_7$

$$\begin{aligned} \dot{X} &= -T^+\Lambda_1^2 TX - aT^+\eta - C^T\Omega_7 CX \\ \dot{\eta} &= TX \end{aligned} \tag{5.133}$$

The positive scalar a is a tuning parameter.

The control law is thus

$$f = -\mathbf{M}T^T\Delta_7^{-2}\left(a\eta + \Lambda_1^2 TX\right) - \mathbf{M}\left(C^T\Omega_7 C + K\right)X - ESa_g \tag{5.134}$$

In order to include the input limitations into the design phase, this control law is adapted with the inclusion of saturations in the feedback law according:

$$f_s = -\mathbf{M}A^T\sigma_2\left[AX + \Gamma X + \sigma_1\left[B\eta\right]\right] - ESa_g \tag{5.135}$$

where

$$\begin{aligned} A &= \Delta_7^{-1}\Lambda_1 T \\ B &= a\Delta_7^{-1}\Lambda_1^{-1}\Lambda_1 \\ \Gamma &= \Lambda_1^{-1}\Delta_7^{-1}T\left(C^T\Omega_7 C + K\right) \end{aligned} \tag{5.136}$$

and the saturation functions are defined as:

$$\sigma[z] = \Sigma z \tag{5.137}$$

where the diagonal matrix Σ is defined by:

$$\begin{aligned} |z_i| < m_i \quad &\Rightarrow \quad \Sigma_i = 1 \\ |z_i| \geq m_i \quad &\Rightarrow \quad \Sigma_i = \frac{m_i}{|z_i|} \end{aligned}$$

with $0 < rwm_i < R$ Let's define the candidate Lyapunov function

$$W_2 = \frac{1}{2}$$

Considering the saturated force input, the velocity derivatives

$$\begin{aligned} \dot{X} &= KX + \mathbf{M}^{-1}\left(ESa_g + f_s\right) \\ \dot{X} &= KX - A^T\sigma_2\left(AX + \Gamma X + \sigma_1\left[B\eta\right]\right) \end{aligned} \tag{5.138}$$

leading to

$$\dot{W}_2 = X^T K X - X A^T \sigma_2 \left(AX + \Gamma X + \sigma_1 \left[B\eta \right] \right)$$

The first term is zero since K is antisymmetric. Introducing $z_1 = B\eta$, $z_2 = AX$, $v_2 = \Gamma X$, this equation may be written as

$$\dot{W}_2 = -z_2^T \sigma_2 \left[z_2 + v_2 + \sigma_1 \left[z_1 \right] \right]$$

If the saturations are chosen such that $R_1 < \frac{r_2}{2}$, then for $|z_2| > \frac{r_2}{2}$, $\dot{W}_2 < 0$ showing that W_2 is a Lyapunov function and thus after a finite time T_2, the variable z_2 will enter the linear zone of its saturation and remain into it. After time T_2, the force input is then

$$f_s = -\mathbf{M} A^T \left(AX + \Gamma X + \sigma_1 \left[B\eta \right] \right) - E S a_g$$

and the velocity derivative is

$$\dot{X} = KX - A^T \left(AX + \Gamma X + \sigma_1 \left[B\eta \right] \right) = A^T \left(AX + \sigma_1 \left[B\eta \right] \right) - C^T \Omega_7 C X$$

Recalling that the overall function

$$W = \frac{1}{2} Y_1^T Y_1 + \frac{1}{2} Y_2^T Y_2$$

Differentiating this equation

$$\dot{W} = (a\eta + 2TX)^T \left(aTX + D\Omega_7 CX + TX \right) - a \left(TX \right)^T TX$$

or

$$\dot{W} = -\left(X_2 z_2 + X_1 z_1 \right)^T \left(X_2 z_2 + X_1 \sigma_1 \left[z_1 \right] \right) - a \left(TX \right)^T TX$$

with $X_1 = \Delta_7 \Lambda_1 = (a+2)^{1/2} \Delta_7$ and $X_2 = \Delta_7 \Lambda_1^{-1} = 2(a+2)^{-1/2} \Delta_7$.

From the above formulas and considering that all the design parameters are positive, it may be verified that both X_1 and X_2 are positive diagonal matrices and their minimum and maximum eigenvalues verify

$$\lambda_{max}(X_2) < \sqrt{2} < \lambda_{min}(X_1)$$

which leads to

$$|z_2| < |\sigma_1 [z_1]| \quad \Rightarrow \quad |X_2 z_2| < |X_1 \sigma_1 [z_1]|$$

Then $\dot{W} < 0$ showing the exponential stability of the controlled system.

5.3.5 Line Tracking by Path Curvature and Torsion

The trajectory $C(s)$ expressed in terms of this coordinate system is in local canonical form

$$
\begin{aligned}
x(s) &= s - \frac{\kappa(s_0)^2 s^3}{6} \\
y(s) &= \frac{\kappa(s_0)s^2}{2} + \frac{\kappa'(s_0)^2 s^3}{6} \\
z(s) &= \frac{\kappa(s_0)\tau(s_0)s^3}{6}
\end{aligned}
\tag{5.139}
$$

In the neighborhood of $C(s_0)$, the curve $C(s)$ has third order contact to a circular helix set at its Frenet-Serret frame

$$C_1(\bar{s}) = \left(a\cos\frac{\bar{s}}{c}, a\sin\frac{\bar{s}}{c}, -\frac{b\bar{s}}{c}\right)^T$$

with

$$a = \frac{\kappa(s_0)}{\kappa^2(s_0) + \tau^2(s_0)}, \qquad b = \frac{\tau(s_0)}{\kappa^2(s_0) + \tau^2(s_0)}, \qquad c = \sqrt{a^2 + b^2}$$

When the vehicle is supposed to track a directed line L, the following control can be proposed

$$
\begin{aligned}
\frac{d\kappa}{ds} &= -a_1\kappa - a_2(\gamma - \gamma_{ref}) - a_3(\chi - \chi_{ref}) - a_4(dist)_{plane} \\
\frac{d\tau}{ds} &= -a_5\tau - a_6(\gamma - \gamma_{ref}) - a_7(\chi - \chi_{ref}) - a_8(dist)_{outside}
\end{aligned}
\tag{5.140}
$$

where a_i are positive constants. This negative feedback rule is called a steering function. The first terms $-a_1\kappa$ or $-a_5\tau$ are feedback terms for damping. The second terms $-a_2(\gamma - \gamma_{ref}) - a_3(\chi - \chi_{ref})$ or $-a_6(\gamma - \gamma_{ref}) - a_7(\chi - \chi_{ref})$ are a feedback terms for the angle errors, finally the last terms are a feedback term for the positional error.

5.3.6 Intelligent Control

Intelligent control as a new trail is also transferred to the LTAR control from the modern robotic research. Intelligent control is a general term that describes a diverse collection of non-traditional control techniques based on soft computing approach [159, 164, 165]. These include neural networks, fuzzy logic, adaptive control, genetic algorithms and several others. Often they are combined with each other as well as with more traditional methods; for example, fuzzy logic controller parameters being optimized using genetic algorithms or a neural network driving a traditional linear controller. Reinforcement learning is one intelligent learning technology which can help facilitate an easier design for autonomous control system and reduces human interaction as much as possible in aerial flights.

Artificial Neural Networks, in analogy to biological structures, take advantage of distributed information processing and their inherent potential for parallel computation. Neural network (NN), very basically, is a network of simple nonlinear processing elements (neurons) which can exhibit complex global behavior determined by element parameters and the connections between the processing elements. It has been shown that a certain class of neural network (NN) can approximate any continuous nonlinear function with any desired accuracy [200]. This property allows to employ NN for system identification purposes, which can be performed both off line and on line. Another area of intensive application of NN is fault tolerant control, made possible due to on line adaptation capability of NN. In [165], an 'add-on' neural controller is developed to increase auto-landing capabilities of a damaged aircraft with one of two stuck control surfaces. A significant increase in successful landings rate is shown, especially when the control surfaces are stuck at large deflections. Fuzzy logic control also gained some popularity among flight control engineers. This type of control relies on approximate reasoning based on a set of rules where intermediate positions between 'false' and 'true' are possible. Fuzzy logic control may be especially useful when a decision must be made between several controversial conditions [87, 94, 112, 200].

Evolutionary and genetic algorithms (EAs, GAs) are global optimization techniques applicable to a broad area of engineering problems. They can be used to optimize the parameters of various control systems, from simple PID controllers to fuzzy logic and neural network driven controllers. Another common design approach is evolutionary optimization of trajectories, accompanied by a suitable tracking controller.

Remark 5.8 The difficulty can be resolved by using a systematic approach based on genetic algorithms, also referred to as evolutionary methods. They do not require any guess for the variables of the problem, the parameters involved in the process are coded as binary arrays and a complete set of them forms an individual. Genetic algorithms do not need any guess because the starting population is randomly generated. However, they require the definition of the search space for all the parameters, that is the ranges in which the values of the parameters are to be searched. Suitable operators, such as crossover and mutation, govern the reproduction mechanism and the population is improved (with reference to the objective function) generation after generation. Elitism is an additional operator that is usually employed to preserve the best individual in the generation. Basically, evolutionary method constitute an effective statistical search technique for selecting the best parameters that is the parameters that minimize the objective function. At the end of the process, genetic algorithms are expected to generate the best individual which includes the optimal values of all the parameters.

Model Predictive Control has been combined with dynamic model analysis for airship control. Ko in [112] presented a new type of Model Predictive Control which was related to the Gaussian processing and reinforcement learning technology.

In [179], the problem of learning to control the height of an autonomous blimp on line and without pre-defined physical models is investigated. The goal is to allow the lighter than air robot to autonomously learn the actions necessary to maintain its height from scratch after it has been switched on. The lighter than air robot should adapt its behavior to the environment and its own dynamics such as the amount of Helium contained in the envelope. The lighter than air robot should actually learn new policies from scratch whenever needed and still being able to fulfill its mission. The approach is based on the reinforcement learning controller, i.e. the goal to be achieved by the controller is specified by virtual rewards given to the system when certain system states are reached. This approach employs a Kalman filter to estimate the ground clearance based on noisy distance measurements obtained from an ultrasound range sensor. Based on these height estimates, Monte Carlo reinforcement learning is applied in combination with Gaussian processes to represent the Q function over the continuous state action space. Reinforcement learning is based on the idea that an agent interacts with a potentially unknown environment and gets rewarded or penalized according to the actions it performs. In general, the agent receives rewards for actions that are beneficial in certain states for achieving a long term goal. The agent thus seeks to behave in a way that maximizes its numerical reward. A reinforcement learning task can be defined by a quadruple $\{S, A, \delta, r\}$ consisting of states S, actions A, a transition function $\delta : S \times A \to S$ and a reward functions $r : S \times A \to \Re$ which defines the immediate reward to be received when executing action a in state s. The goal is to define a policy $\Pi : S \to A$ which maps each state s to an action a, such that the future expected reward is maximized. The expected long time reward in state s_t can be expressed recursively as $R^{\Pi}(s_t) = \sum_{i=0}^{\infty} \varsigma^i r_{t+i}$ where $\varsigma \in [0, 1]$ is a discount factor. In general, the sequence of rewards r_{t+i} is obtained by starting in state s_t and then iteratively applying a policy Π for selecting the next action

$$a_t = \Pi(s_t)$$

Several approaches may be proposed to solve the reinforcement learning problem by maximizing the expected long time reward. In [179], the Monte Carlo method, which has the advantage that it allows the agent to learn directly from experience while interacting on line with a completely unknown environment. This enables the lighter than air robot to learn without prior knowledge and also in situations in which no simulation environment is available.

5.4 System Health Management

System Health Management is considered as a system engineering discipline that includes the processes, techniques and technologies used to design, analyze, build, verify and operate a system to prevent faults and to minimize their effects. System Health Management, when implemented successfully greatly enhances safety, affordability and maintainability of complex systems [117].

A System Health Management system consists of instrumentation components (sensors, wires, data recorders...) a Fault Detection Isolation and Response (FDIR) module, diagnostic and prognostic software as well as processes and procedures responsible for information gathering about a system's health and corresponding decision making. SHM are typically used to accomplish two main goals:

- to accurately assess the system health and to pinpoint problems and anomalies, monitoring the health of a system and detecting and isolating faults and responding to potential problems by enabling system reconfiguration or restoration.
- to support the management of the system's off-line maintenance and repair operations, by mapping anomalies to physical components that have failed during operation and by predicting physical components that are likely to fail in the future. These components, referred as Line Replaceable Units (LRU) are then replaced to restore a failed system function.

Many aspects of System Health Management have been implemented in many systems. What is needed is a coherent framework to integrate across instrumentation design, on board Fault Detection Isolation and Response, ground maintenance, diagnostics and various fault-related analyzes applied to lighter than air robots. A systematic design methodology, Functional Fault Analysis (FFA) can be developed with the goal of integrating System Health Management into early design of complex systems.

The basis for the Functional Fault Analysis methodology is a high level functional model of the system that captures the physical architecture, including the physical connectivity of energy, material and data flows. The model also contains all sensory information, failure modes associated with each component of the system, the propagation of the effects of these failures modes and the timing by which fault effects propagate along the modeled physical paths. Once, this integrated model is built, the designers and system analysts can assess the capability of the sensor suite of the system to isolate the location of faults, and determine if redundant sensors exist to confirm the existence of a fault. Moreover, these capabilities can be used to assess the sensor suite's diagnostic functionality and to analyze the race between the propagation of faults effects and the Fault Detection Isolation and Response mechanisms designed to compensate and respond to them.

The Functional Fault Analysis methodology offers advantages for both the design and the operational phases. During the design phase, the Functional Fault Analysis provides the ability to:

- Assess the effectiveness of the sensor suite to isolate faults to the Line Replaceable Units.
- Model the fault effect propagation paths and assess the time latencies along those paths.
- Document and analyze the Fault Detection Isolation and Response time response capability in terms of sensor detection capability, sensor confirmation, and the time from fault initiation until detection is confirmed.

- Assist design engineers by uncovering problems in design issues across subsystem boundaries.

Failure Modes and Effects Analysis (FMEA) has been the most widely used technique for conducting fault impact analysis. This technique systematically examines individual system components to assess risk and reliability. It is a bottom-up approach that starts at the component level and follows an inductive logic to determine the consequences of critical component failures. Accordingly, failure mechanisms in which each component can potentially fail are identified and evaluated separately to determine what effect they have at the system level. Complementary to Failure Modes and Effects Analysis, Fault Tree Analysis (FTA) is performed using a top-down approach. Fault Tree Analysis starts with identification of a high level failure event. A deductive logic is then followed to drive contributing events that could lead to the occurrence of immediate higher level events. At the end, the analysis presents the chain of events combined with logical gates in a tree structure. Using this approach, possible event paths from failure root causes to top level consequences can be captured. Finally, Probabilistic Risk Assessment (PRA) is a method that combines a number of fault/event modeling such as Master Logic Diagrams (MLD), Event Sequence Diagrams (ESD) and fault trees and integrates them into a probabilistic framework to prioritize risk drivers during design.

5.4.1 Health Monitoring

The need for increased flight safety leads to the design of reconfigurable fault-tolerant control system. The main objective of fault tolerant control is to maintain the specified performance of a system in the presence of faults. Two approaches can be distinguished in this area: the passive and the active approaches. In the passive approach, the control algorithm is designed so that the system is able to achieve its given objectives, in healthy as well as in faulty situations. Unfortunately achieving robustness to certain faults is only possible at the expense of decreased nominal performance. The active approaches react actively to fault events by using reconfiguration mechanisms. Consequently, this ensures nominal performances in fault-free situations. An active fault tolerant control is characterized by on line and real time fault detection and isolation and a reconfiguration mechanism. This scheme requires its control law to react to faults through reconfiguration and Fault Detection and Isolation (FDI) modules. The goal is to maintain overall system stability and acceptable performance in spite of the occurrence of faults by reconfiguring the nominal control law when a fault is detected by the FDI unit. The FDI mechanism is supposed to detect and diagnose any relevant failure that could lead to flight performance degradation. This shall be done sufficiently early and in compliance with the stringent operational and flight dynamic constraints, to set up timely safe recovery actions and to improve the situation awareness of the flight management system.

One approach to detect and isolate actuator or sensor faults is the Multiple Model Adaptive Estimation (MMAE) methods. It is based on a bank of Kalman filters (KF), each of which is matching a particular fault statics of the system. A hypothesis testing algorithm uses the residuals from each Kalman filter to assign a conditional probability to each fault hypothesis. The main advantage of the method lies in its responsiveness to parameter variations, leading to faster fault isolation than that attained by other methods without multiple—model structure. It also enables the reconstruction of a correct state estimate even when an actuator or sensor fault occurs, because the estimated state vector is the sum of each KF estimate weighted by its corresponding probability. The multiple model adaptive estimation method can be applied in practice as long as the expected faults can be hypothesized by a reasonable number of Kalman filters. To make the multiple model adaptive estimation method applicable for any flight condition and capable of isolating lock in place or floating actuator faults, the multiple model adaptive estimation algorithm is combined with Extended Kalman Filter (EKF) used for the nonlinear estimation of some (unknown) fault parameter: the deflection of a faulty control surface or actuator.

A fault tolerant control system needs to first detect and isolate the presence and location of faults, and then recover from the identified faults by reconfiguring the control system architecture. The development of these three tasks is known as the Fault Detection, Isolation and Recovery (FDIR) problem. An extensive body of work has been conducted on FDIR design of linear systems. Among them the parity space approach which basically performs consistency checks on the mathematical equations governing the system has been studied [30, 117, 146].

Besides the parity space approach, many researchers simply use Luenberger observers or Kalman filters to monitor system performance. In the absence of modeling uncertainties and disturbances, the errors between the state estimates and the actual states, namely the residuals, diminish to zero, whereas the presence of a fault causes the residuals to deviate from zero. In order to avoid false alarms caused by uncertainties and disturbances, the threshold must be sufficiently larger than zero, although this will subsequently decrease the sensitivity of the detection system to lower magnitude faults.

The methods described above are applicable to the linear approximation of the lighter than air robot model.

5.4.2 Diagnosis, Response to Systems Failure

When a classical airship has a failure in flight, for example total loss of power or the flying control system, it can revert to being a free balloon. When the vehicle is hovering without any payload and low on fuel, it is light enough to maintain altitude and land safely in the event of power failure in one of the rotors. A particular concern is that off-nominal conditions or degraded components could reduce the capabilities of the lighter than air robot to accomplish the mission objectives. A fault is defined as any kind of malfunction in the actual physical system that tends to degrade the overall system performance. Such malfunction may occur either in the sensors (instruments), actuators or in the components of the process. To maintain a high level of performance, it is important that failures can be promptly detected and identified so that appropriate remedies can be applied.

In the past decades, numerous approaches to the problem of failure detection, isolation and accommodation (FDIA) in dynamic systems have been developed. Among them are two major philosophies: physical redundancy and analytical redundancy. Physical redundancy is achieved simply through hardware replication. Unlike physical redundancy, analytical redundancy is a model-based method. Analytical redundancy methods have many advantages over physical redundancy methods: the replication of identical hardware components (actuator/sensor) is more expensive, restricted and sometimes difficult to implement in practice. Fault detection and isolation can be achieved by the eigenstructure assignment method and Linear Matrix Inequality technique.

There has been significant research on the observer approach, the parity space approach and the robust parameter estimation approach. Parity equations are generally constructed to detect the fault and isolate the faulty sensor. The value of the parity equation related to the faulty sensor has a non zero value over the threshold and therefore a faulty sensor can be identified by comparing all the values of the parity equations.

Especially, eigenstructure assignment is realized by the concept of detection spaces. Once a set of stable eigenvalues with respect to each independent detection space is assigned, the LMI technique can be used to find the optimal detection filter, so that the noise effect on the residual is reduced. The proposed filter has good performance as well as the following advantages:

- The eigenvalues are assigned in the detection space, which is the sum of minimum (C, A) invariant subspace and invariant zero directions for F_i. It extends the applicability of the detection filter.
- The designed filter is robust to disturbances, which makes the filter more practically useful
- The LMI technique is used to solve the optimal filter problem, which is computationally efficient. The proposed method can be used to perform real-time detection and isolation.

The state-space model of the linear time invariant model with additive faults and without disturbances in continuous time format, approximation of the lighter than air robot, is

$$\begin{aligned} \dot{X}(t) &= AX(t) + BU(t) + \sum_{i=1}^{q} F_i m_i(t) \\ Y(t) &= CX(t) \end{aligned} \tag{5.141}$$

When no faults occur, $m_i(t) = 0$. The fault direction F_i can be used to model actuator, sensor and components faults. The system is assumed to be observable. Considering the following full order observer, the following error system dynamics is obtained

$$\dot{\varepsilon}(t) = (A + LC)\varepsilon(t) - \sum_{i=1}^{q} F_i m_i(t)$$
$$r(t) = C\varepsilon(t) \tag{5.142}$$

where $r(t)$ is the residual output. Clearly, if L is chosen such that $A + LC$ is stable, $r(t)$ in the steady state will deviate from zero when any faults has occurred. But it is not enough because different faults must be isolated and that is a more difficult task. To detect and isolate faults, a set of detection subspaces $T_1, T_2, \dots, T_q$ is assumed to satisfy the following three conditions:

$$(A + LC)T_i \subseteq T_i$$
$$F_i \subseteq T_i \tag{5.143}$$
$$CT_i \cap \sum_{j \neq i}^{q} CT_j = \varnothing$$

Then the residual $r(t)$ has a unique representation of $r(t) = z_1 + z_2 + \dots + z_q$ with $z_i \in CT_i$. Therefore, faults can be isolated by projecting $r(t)$ onto independent subspaces CT_i. The detection subspace T_i can be spanned by the following linearly independent vectors:

$$T_i = \begin{bmatrix} \nu_{is} & \dots & \nu_{i1} & F_i & AF_i & \dots & A^{k_i} F_i \end{bmatrix} \tag{5.144}$$

where k_i si the smallest integer such that $CA^{k_i} F_i \neq 0$ and z_{ik}, ν_{ik} are the invariant zero and associated zero direction, which are defined as $k = 1, 2, \dots, s$

$$(A + F_i K)\nu_{ik} = z_{ik}\nu_{ik}$$
$$C\nu_{ik} = 0 \tag{5.145}$$

Because $(A + LC)T_i \subseteq T_i$ and $Ak_i F$ is in the column space of T_i, we can define

$$(A + LC)\, Ak_i F_i = T_i x_i = T_i \begin{bmatrix} \beta_{is} & \dots & \beta_{i1} & \alpha_0 & \alpha_1 & \dots & \alpha_{k_i} \end{bmatrix} \tag{5.146}$$

5.5 Conclusions

The lighter than air robot is a nonlinear system, but for classical controller design, linearized models obtained at some operating points are used in aerodynamic flight. In practice, it is necessary to determine linear models at several design operating points over a specified flight envelope and determine optimal control gains for each one. Then the design control gains are tabulated and scheduled using microprocessors, so that the gains most appropriate for the actual operating points of the lighter than air robot are used in the controller. It is usual to determine which of the design operating points are closest to the actual operating point and use a linear combination of the control gains corresponding to these control points.

As shown in the second part of the chapter, dynamic response at the hover is characterized by the stability modes which in turn are determined by the aerodynamic properties. Since the aerodynamic contribution is negligibly small both the longitudinal and lateral non-oscillatory stability modes have time constants measured in hours than seconds. Similarly, the oscillatory stability modes have negligible aerodynamic damping. For all practical purposes, the lighter than air robot appears to be neutrally stable at the hover.

As shown in the third section of the chapter, airships offer a control challenge as they have non zero drift and their linearization at zero velocity is not controllable. Several nonlinear control techniques have been used such as dynamic inversion, variable structure control. Finally, system health management is discussed at the end of the chapter.

Chapter 6
General Conclusions

Lighter than air systems are particularly appealing, since the energy required to keep them airborne is small. The Lift of the lighter than air robot is mainly aerostatic, as opposed to aerodynamics as in airplanes and helicopters. Consequently, lighter than air robots spend most energy moving and compensating wind disturbances, rather than trying to keep themselves on air. Guidance and control methods of such vehicles have been presented throughout this book. However, research is very active in this field and many challenges still exist.

The challenges faced by the Lighter Than Air Robots (LTAR) span several and distinct fields, including guidance, control, navigation, collision prevention, take-off/landing techniques, state regulations, safety/reliability, man-machine interface design issues...

1. The ability for a vehicle to manage its position away from obstacles represents a significant issue and a necessity for low altitude operations in crowded environments or their ability to detect hard to see obstacles such as suspended cables quickly. All operations involving close coordination and physical interaction between vehicles or between vehicle and the ground require further research. Flight in cluttered environments includes any phase of the flight where the vehicle is in close proximity to obstacles. This flight mode is particularly important for low altitude applications, using a variety of sensing techniques. Low altitude aerial robots, often operating in a cluttered environment, offer the opportunity to explore many generic robotics topics, including vision, path planning, mapping and other algorithms in a progressive manner.
2. Another challenge of autonomously controlling/navigating the airship from low to high airspeeds, for example, from vertical take-off/landing and hovering flight up to aerodynamic flight. Given such a broad operational range, vehicle control mainly varies from propulsion and propulsion vectoring (at low speeds) to propulsion and aerodynamic surfaces (at high speeds), stressing the abrupt change and non-linear behavior of airship dynamics. Therefore, complex non-linear control techniques for airship control and guidance are under investigation, including atmospheric disturbances such as wind and gusts.

Y. Bestaoui Sebbane, *Lighter than Air Robots*,
Intelligent Systems, Control and Automation: Science and Engineering 58,
DOI 10.1007/978-94-007-2663-5_6, © Springer Science+Business Media B.V. 2012

3. Decision making in reality always consists of uncertainty. Uncertainty is not an occasional, temporary occurrence in decision planning. A realistic and robust decision cannot be made without understanding and bringing uncertainty into the planning process. Under uncertainty, what is required is a decision that performs well over all the possible scenarios. This is known as the robust decision making.
4. Future LTAR will be operated with algorithms capable of monitoring the airship health and of taking action if needed. Fault tolerant control systems for small and low-cost LTAR should not significantly increase the number of actuators or sensors to achieve the safer operation.
5. One of the most difficult issues related to autonomy is operating in a highly dynamic environment with other vehicles operated by humans. Humans are capable of generating some highly unpredictable behaviors therefore an autonomous vehicle must be flexible enough to deal with unpredictable events or situations.
6. Last but not least, all the algorithms must be implemented on real-time microcontrollers.

Appendix
Current Projects

A.1 Introduction

After being neglected for a few decades, airships are now again a source of high interest, and new research programs are being launched around the world. This new trend in favor of airships is mainly due to at least three reasons: a major concern for sustainable growth, an increasing need for carrying heavy loads, and new expectations for survey and monitoring means. Their use represents a niche in the aeronautical market.

Operational airships in the world are usually propelled by fossil combustible engines. Only small airships, such as robotic airships, are electrically propelled. One goal is to imagine and validate the concept of an airship using only non-fossil and renewable energy, with the highest level of autonomy. This goal has to be achieved with the highly airship-specific constraint of weight minimization. Effectively, each increase in weight results in an increase in gas volume required for sustaining the overall airship load, typically roughly 1 m^3 for any 1 kg in load increase. Two sources of energy are been studied: solar energy retrieved on the upper surface of the airship envelope, and fuel cell energy. The large uppers surfaces of airships could provide with an appropriate solar exposure but, due to their convex shape, imply the use of flexible photovoltaic cell arrays [26].

The projects presented below have the advantage of coupling academic research with development of a vehicle platform. Another characteristic is the interdisciplinary aspect. Disciplines involved are aerodynamics, flight mechanics and control, avionics and navigation, mission planning and replanning, structure, power systems, computing, embedded systems, mechatronics and interdisciplinary design aspects, not forgetting the societal aspect.

A.2 Artic Airship

Advances in technology make airships one of the most promising modes of freight transport for the 21st century. Interest has been growing in the potential for air-

Y. Bestaoui Sebbane, *Lighter than Air Robots*,
Intelligent Systems, Control and Automation: Science and Engineering 58,
DOI 10.1007/978-94-007-2663-5, © Springer Science+Business Media B.V. 2012

ships to serve remote areas that have fewer transportation alternatives and scarce infrastructure. Historical precedence suggests that airships can operate safely in cold weathers. These locations hold the promise to begin the explosive growth in demand for Lighter than air technology. No existing buoyancy aircraft is certified for operations in icing conditions. Consequently, the first assumption in formulating a qualitative forecast is that new airships can be engineered to operate safely in winter cold, precipitation and wind conditions. The only infrastructure requirement is an unimproved runway or expanse of lake for takeoff and landing. Hybrid aircraft could become the dominant mode of transport in circumpolar regions [173].

For mining demand, mines are operated by a combination of winter roads and small aircrafts. Winter roads are used to bring in the year's supply of diesel and heating oil to operate the mines. Remote communities in circumpolar regions depend on annual sea lift, ice roads and small airplanes for transportation services. Annual services transport the heavy indivisible role for hybrid aircraft or bulky goods but are inconvenient.

A particular concept that belongs to the class of vertical takeoff and landing aircraft with unprecedented lifting capability is the quad-rotor hybrid airship [157]. Basically, this concept consists of a non rigid buoyant non rotating hull that is rigidly attached to a structural frame supporting the propulsion components. This vehicle consists of a buoyant envelope attached to a supporting structure to which four rotor systems are attached. The rotors are used for control and maneuvering in near-hovering flight. The form of this airship is not classical too. This airship has a delta wing form, which compensates for a considerable part of the total weight of the vehicle with aerodynamic lift. The forces and moments required to control the aircraft with or without its load are obtained by changing the lifting rotor thrust vectors as well as thrust generated by the auxiliary propellers, in unison or differentially. The empennage control surfaces provide additional control moments in pitch and yaw during forward flight. The idea of combining aerodynamic and aerostatic lift is not a new one. The first patent for a 'semi buoyant aircraft' arose in 1931. Later, hybrids with high aspect ratio wings, lifting bodies or those operating in ground effect were proposed. In addition, a concept has appeared in the combination of helicopter and airship. The advantage of such arrangement is that the empty weight of the vehicle is supported by the force due to buoyancy while the propulsive forces are entirely available for lifting the payload and controlling the vehicle. This complexity is worthwhile, given the objective to provide a large payload, high manoeuvrability and operation costs well below that of a comparable helicopter. The main idea of this proposal is to combine both aerodynamic lift with quad-rotor structure, making this airship hybrid in both ways.

Many types of emergency response can be envisioned but oil spills and forest fires are the most pressing problems in the North. Hybrid aircraft could provide communications and fire intelligence to manage the emergency response. Two key objectives for this type of mission are that the airship has exceptionally long endurance and that it operates with a sufficiently high level of autonomy, to lighten the heavy workload of the operator. In order to achieve these objectives, a robust guidance and control system is required, capable of auto piloting and controlling the airship under an extremely wide range of atmospheric and wind conditions.

The objectives of the project are the following:

1. A small-scale flight research vehicle must be designed and built. Its maneuverability and handling qualities have to be investigated at low airspeeds and Stability characteristics analyzed.
2. Nonlinear equations of motion capable of modeling the dynamics of this coupled multi-rotor/support frame have to be developed. Using these equations, an aeromechanical stability analysis has to be performed in order to identify potential instabilities of this type of vehicle. The coupling between various blade supporting structure and rigid body modes have be identified. The effects of changes in buoyancy ratio (buoyant lift/total weight) on the dynamic characteristics of the vehicle have to be studied. Mechanical and aerodynamic coefficients must be identified.
3. Flight dynamics simulator.
4. Trim characteristic of the vehicle have to be discussed in hovering flight and in normal flight.
5. Navigation and Path planner with realistic wind information. How to use meteorological information into the flight management system. Mission reconfiguration and trajectory optimization.
6. Sensor and actuator suite, advanced communication.
7. Choice of the control platform software architecture.
8. Concepts for controlling the vehicle with or without its external load have to be evaluated.
9. Response to control inputs while hovering over a point have to be shown. Effects of the varying mass due to fuel consumption must be analyzed.
10. Effects of atmospheric turbulence in flight or in hover should be investigated.
11. Diagnosis, fault-tolerance and systems failure must be investigated.
12. Last but not least, Feasibility study in ground effect and Long range and endurance power and propulsion have to be checked.

A.2.1 Vehicle Description

Based on a preliminary design study, a vehicle configuration is selected that consists of four modified rotors mounted on an interconnecting structure that is attached to a delta wing airship envelope [26]. In order to determine the effect of vehicle buoyancy on its flight characteristics, it is designed to be closer to neutral buoyancy while it is empty and with maximum payload. Heaviness of the airship corresponds to gross weight in excess of static lift for that flight condition. This delta wing form compensates for a considerable part of the total weight of the vehicle with aerodynamic lift. The forces and moments required to control the aircraft with or without its load are obtained by changing the lifting rotor thrust vectors as well as thrust generated by the auxiliary propellers, in unison or differentially. The empennage control surfaces provide additional control moments in pitch and yaw during forward flight. For the present vehicle configuration, control variables may be systematically combined for controlling the airship in all of its flight modes, including precision hover. A particular choice of these options, in a sequence or in unison, depends on the flight and operating conditions. Clearly, in the absence of a payload the airship is light enough that very little rotor thrust is required to lift and sustain it in flight. Consequently, in this case, rotor thrust vectoring for controlling the vehicle is ineffective. However, the auxiliary propellers can be conveniently used in this mode to control the vehicle. These constraints suggest that in a case where the airship is hovering over a point on the ground, the lifting rotors should be used for vehicle control when it is heavy and the auxiliary propellers when it is light. In the ferry mode when the vehicle is not carrying any payload it may be controlled like a conventional airship, particularly during short takeoff and landing while the lifting rotors may be used for control augmentation only. An important aspect of the vehicle control configuration is selecting the location of its auxiliary propeller components. Typically, the line of action of the resultant of thrust vectors determine the vehicle attitude and its trimmability during hover. The vertical location of these components must be selected on the basis of operational and ground handling considerations [55].

A.2.2 Weight, Mass Distribution and Balance

Weight, mass distribution and balance are difficult to achieve, since small tolerance differences over such big structures have a significant effect. For example, if actual fin weights do not match their design estimates overall balance will be affected owing to the fin's extreme position. It is also difficult to predict accurately the position of the center of lift. This is because the gas container varies accordingly to applied pressures and other loads, its materials of construction and inertia and gravitational effects. For airships, it is the relationship between the center of gravity and center of lift which is the important parameter. For the airship to remain statically level, the center of gravity should be directly below the center of lift. Any horizontal offset will result in the airship adopting a pitch angle. The vertical separation between the center of gravity and the center of lift affects the handling characteristics of the airship, the greater the separation the more stable the airship becomes, though greater control inputs are required are required for pitch and roll maneuvers. During the design phase, much effort should be spent ensuring that the center of gravity is as close as possible to the optimum position when considering the loaded airship [108].

A.2.3 Modeling and Identification

Complete rigid body motion of the airship has to be simulated using general non linear equations of motion. Basically, the airship hull is assumed in first approximation, to be a rigid body supporting the gondola and the fins. External forces and moments acting on the vehicle due to gravity, lift, aerodynamics and control inputs must be included in the model. Similarly, acceleration dependent aerodynamics forces and moments on the airship have to be estimated and included [36].

A.2.4 Aerodynamics

The aerodynamics of an airship shows fundamental differences with the airplane aerodynamics. Whereas airplane wings are relatively thin and of high aspect ratio, the envisaged deltoid wing is necessarily thick, in order to create lift, and of small aspect ratio. Of course, the results obtained on Delta wings which generally equip fighter airplane cannot be used for this deltoid wing due to differences in velocity, leading edge shape and thickness. But above all, a more fundamental difference concerns the behavior in unsteady movements of the body. Due to the fact that a airship has its own mass M of the same order of magnitude than the mass M_a corresponding to lift, all the unsteady aerodynamic coefficients, proportional to M_a are of considerable importance whereas they are negligible for a heavier than air airplane. The consequence is that the mass and inertial characteristics of an airship are increased by the presence of the surrounding air, and that this increase is different according to the direction of the acceleration and their knowledge are of major necessity for a flight mechanics model. The final aim of the aerodynamics study is to built a complete model for the aerodynamics of the deltoid wing, including parameters like velocity, angle of attack, side slip angle, roll, pitch, yaw but also all the derivatives with respect to time of these parameters. The means to provide in order to achieve this goal are:

- Wind tunnel tests for the steady aerodynamics.
- Dynamic tests on a flying mock-up in order to find the unsteady aerodynamics coefficients by the analysis of the trajectories.
- Computational fluid dynamics to compare with wind tunnel tests and assess the capacity of the numerical codes for prediction of the aerodynamic coefficients.
- Theoretical study to establish the basis of the model either for steady or unsteady flow.

A.2.5 Localization and Positioning

Localization of the vehicle is the first step for an autonomous navigation system. Accurate localization information is also needed by the geographical (in 3D) and temporal mapping of the on-board experimental data. Autonomous positioning can be achieved by exploiting navigation satellite data (GPS, GNSS), and by means of semi-passive back-up techniques. Classical aircraft on-board instruments are also provided to the pilot in case of failure of the autonomous system. There are many absolute localization (and navigation) devices using vision systems, artificial beacons, ultrasonic or laser range finders, and radio-systems. The satellite navigation allows to obtain global positioning in the reference system fixed to the Earth. They provide medium or high independent accuracy of positioning. An enhanced system will use redundancy and complementarity of data delivered by several localization sources of different natures. Specific data fusion and interpretation processes must then be studied, depending of the chosen multi-sensor system and the desired accuracy and reliability.

The challenge for lighter than air robot today, is to increase their level of autonomy. One important aspect of autonomy is their capacity to automatically construct a map of its environment starting in a partially known or unknown setting and simultaneously localize itself within this map. This airborne simultaneous localization and mapping (A-SLAM) problem remains a very important problem to address, as its solution can postulate as a strong candidate to achieve seamless navigation. Classically, GPS information is used to correct drifts of the inertial sensors on board the autonomous vehicles. However, GPS is not always available, due to poor or absent GPS coverage.

One of the popular approaches proposed to solve the simultaneous localization and mapping (SLAM) problem uses Kalman Filtering techniques. Extensive research works have been reported in the literature, employing the Extended Kalman Filter (EKF) to address the SLAM problem. Techniques based on particle filtering schemes and unscented Kalman filtering scheme are also investigated. An EKF based SLAM approach estimates and stores the vehicle pose and the map feature positions in one state vector. The uncertainties of these state estimates are stored in an error-covariance matrix that includes cross-correlations between feature/vehicle pose estimates.

In A-SLAM, optical cameras represent an interesting way to providing vast or precise information content. The acquired images are affected by many geometrical and photo-metrical transformations. The 6 degrees of freedom that characterize the movement of the aerial vehicle lead to different geometrical transformations between an acquired image and the next acquired one. In addition as the aerial vehicle navigates in a natural environment, the acquired images often present luminosity and contrast changes [202, 211].

A.2.6 Navigation and Path Planner

The navigation subsystem is very important for aerial vehicles. It is necessary to use absolute navigation and positioning that allows to obtain global and time-independent information. Inertial measurements are essential for stability control, but navigation and other high level tasks demand an additional global localization system. A typical solution adopted for vehicles is the use of GPS, but the precision and the availability can appear too poor in particular flight configurations, as well as the time of delivery and processing too long. Other systems are then necessary to insure an accurate navigation at full time, coupled to the localization device. To navigate accurately, series of predefined way-points can be delivered, either by aeronautic active radio devices or by extraction of particular objects of the environment. Both these solutions seem inefficient in the arctic area. Suitable other techniques must be studied to plan the route in order to deliver information to the path-planner.

This operation must handle meteorological information, especially force and direction of the wind, in order to plan trajectory corrections but also for a secure auto-piloting.

A.2.7 Feeding the Path Planner with Realistic Wind Information

Flight management of an vehicle in a dynamic partially known environment is a complex real time problem. The flight route can be preprogrammed at launch based on current weather conditions and forecast information, but once en route, changes to the flight plan have to be made in accordance to changes in the weather. Even though forecast models have improved, weather and wind patterns continuously change and are still difficult to predict. Thus, the use of frequently updated weather information during the flight will allow adaptation and replanning based on current and near-term forecasts. The Weather forecast must become an important part of the auto-pilot and the flight management system, to detect signs of a slight change in wind velocity, monitor development and movement of clouds, recognize conditions suitable for the development of temperature inversions or of local wind phenomena.

A.2.8 Data Processing and Transmission

In the field of transmissions and telecommunication, difficulties come from the high data flow used for the exchanges with a distant survey or data collecting platform. Both the abroad scientific instruments and the autopilot and the flight management system (including localization and navigation supports) need communication services. For this purpose, intelligent algorithms for on-board processing of data are necessary, in particular the matching of data provided by sensors, and also regarding to their position and timing. Theses processes can include image registration, resolution enhancement, motion-blur compensation, segmentation, feature detection. Some data fusion processes can enhance the accuracy and reliability of the data. Others can extract information of higher level, as symbolic ones, giving a more relevant and compact package of knowing than the rough data. Combined with compression algorithms these processes are then able to comply with the limited bandwidth of the downlink, so that only the most important data are selected and transmitted.

A.2.9 Airship Piloting and Response to Wind Disturbances

The differences on piloting an airship from that of fixed or rotary wing flight includes lower speeds and slower responses and the possibly greater impact of certain meteorological conditions. Every landing is different owing to a combination of meteorological, local and on-board factors: trim slightly different; static heaviness altered by superheat effects; wind velocity slightly varied; different eddy currents or thermals; slightly altered flight path, descent rate or ground speed. The payload of airships is generally limited by the gas lift available in the climatic conditions prevailing at the cruise flight altitude. This means that the key cruise altitude, temperature conditions, payload and mission profile must be established very early in the design process, as alterations to these can cause big changes to volume. Many of the airship's principal operational problems are associated with the functions of lift control, and of interface movements, i.e. landing, load exchange, and take-off. One of the principal operational requirements of this airship is that it should be able to hover in windy environment. Consequently, the operational flight envelope within which the vehicle, with or without payload can be trimmed in hover has to be determined by considering various wind magnitude and directions. The capability of this research aircraft to perform under adverse weather conditions depends on the adequacy of its control for satisfactory operation. It is convenient to consider various operational flight conditions involving step changes in wind magnitude and shifting wind direction to illustrate typical performance of the aircraft [221].

Airships with their nearly neutral lift, large dimensions and relatively low cruise speeds, are sensitive to large scale atmospheric gradients and accelerations. Control concepts must be designed, which emphasize requirements of precision vehicle control and gust response suppression in low-speed and hover flight operations. The utility of this vehicle will depend on its ability to operate in the arctic zone weather. The effects of closed loop control on the hovering performance of this quad rotor hybrid airship in crosswind flight conditions, must be analyzed. The capabilities of this airship to perform under adverse weather conditions would depend upon the adequacy of its control for satisfactory operation. In order to define such a control system and the operational limits of the vehicle itself, it is necessary to characterize the operational flight conditions by representative weather scenarios through measurements and modeling of the atmosphere. Typical atmospheric conditions in the arctic zone, that are critical to the performance of the vehicle have to be identified [2].

A.2.10 Loading and Unloading Lifts

The mass of an airship can vary considerably according to the value of the payload. This complicates the handling of a conventional airship both on the ground and in the air. For the quad rotor hybrid airship, the rotors are not only used for low speed control but also to provide overall lift to compensate for the payload. Recovery on return to the ground provides the most demanding of the ground handling problems. The reason lies in the need to reduce to zero both the horizontal and vertical energy associated with the approach to landing. The problem is the susceptibility of the aircraft to wind variations. The accurate positioning of the airship for the purpose of lifting or placing a load, is difficult. During ground handling, the forward speed of the airship is low in comparison with likely wind speed. The aerodynamic effects resulting from winds are partly reacted by the constraints imposed by handling and partly by inertial effects as the airship responds to the applied forces. In cruising flight, the aerodynamic disturbing effects are reacted to by natural stability, control and inertial effects, so the conditions are quite dissimilar. The problem of controlling the airship alone while it is hovering over a point on the ground in a turbulent cross wind is different from that of the airship with a sling load. In the latter case, it is intended to control the payload motion such that the payload excursions relative to the ground is minimal rather than of the vehicle. The length of the suspension cable is an important parameter in determining the extent of dynamic coupling between the vehicle and payload as well as in designing a closed loop control system for improved vehicle performance. Mathematical models describing the combined motion as well as decoupled motions of the vehicle and payload must be used to determine the effect of motion of the payload suspension point on the payload dynamics. Subsequently, a closed loop control system has to be used to examine the desirable characteristics of such a system in limiting the response of the vehicle, with or without a suspended payload. In order to assess the vehicle/payload system sensitivity to atmospheric disturbances, the turbulence scale length should be arbitrarily varied. Further studies should consider a power spectrum of the wind turbulence model that includes wind acceleration contribution as well. This is of special significance to this vehicle because of its buoyant characteristics. The effects of cross coupling with other lateral and directional dynamics on the system response to cross winds should be investigated. The corresponding model for wind disturbance should include correlated wind turbulence components in all of the rigid body degrees of freedom of the vehicle. The design of the closed loop control system for this system with these constraints is a challenging task [5].

A.2.11 Diagnosis, Response to Systems Failure

When a classical airship has a failure in flight, for example total loss of power or the flying control system, it can revert to being a free balloon. Since the quad rotor hybrid airship is a unique experimental aircraft, its failure modes need a thorough investigation both on ground and in flight. The hybrid nature of the vehicle lends itself to a combination of emergency modes of operation associated with system failures in a delta-wing airship and helicopter. When the vehicle is hovering without any payload and low on fuel, it is light enough to maintain altitude and land safely in the event of power failure in one of the rotors [146].

A.2.12 Flight Dynamics Simulator

Due to the unique rotor configuration and the delta-wing aerodynamic lift, conventional models and tools for rotor crafts dynamics analysis cannot be used. A flight dynamics simulation has to be developed for hybrid heavy lift airships to simulate the present aircraft configuration. Complete six degrees of freedom has to be simulated by using estimated values of the overall physical and aerodynamic properties of the aircraft. Unlike the airplane or helicopter, the handling qualities criteria for hybrid airships are nebulous. Consequently, the control characteristics of the vehicle have to be evaluated by considering specific tasks such as ability to maneuver from hover, ability to accelerate into a heavy wind or cross wind, and ability to hover a point on the ground in a variable, shifting wind. Subsequently, the control power characteristics of the vehicle with or without payload are to be determined by considering the proposed control concepts. Several failure modes of the vehicle are to be simulated to determine their consequence on the flight of the vehicle and its safety. Pilot-in-the-loop simulation will give better insight in this regard, especially if the thrust vectors are to be operated independently [77].

A.2.13 Small Scale Delta-Wing Quad-Rotor Airship

Since the quad-rotor delta-wing airship is a novel vehicle concept, it is proposed to design and build a small-scale flight research vehicle for ground and flight tests to prove the feasibility of the concept and to investigate its flying qualities. For the chosen vehicle configuration, concepts for controlling the vehicle in all of its flight modes have to be developed and evaluated by using a flight dynamics simulation of the vehicle. The response of the vehicle to control inputs and atmospheric disturbances are to be predicted in terms of typical flight parameters. A critical issue for heavy-load airship is the ability to control the altitude of the airship as the load is being secured to the airship structure, carried and finally removed [60, 62, 79].

A.2.14 Ground Handling

The issue of Ground handling is still seen as one of the most problematic areas of airship operation. The solution of ground handling requirements is associated with the provision of adequate control to enable precise low speed flight to be undertaken reliably and safely [55].

A.3 Bridge Monitoring

An important application, bridge monitoring, is highlighted in this section [19, 21]. A flight simulator software is developed to test the various methods of guidance and control proposed in this book. The objective of this section is to present a software simulation of missions to demonstrate innovative concepts for airborne systems. This flight simulator software is used to accurately represent the system

Fig. A.1 Chevire bridge (Nantes, France)

behavior and to generate multiple numerical data useful for analysis of results and algorithms improvement. Thus it is interesting to see a graphical representation of the displacement of the object simulated. Furthermore, the development of a standardized interface between software modeling and visualization environments can more easily use a variety of tools. This flight simulator implements the numerical models: the model of the airship, the simulated models of the environment, the atmosphere and its evolution, the databases needed for scenarios, and finally tools for recording simulation data. This flight simulator software enables to present an easy and understandable way of monitoring mission specialist. The environment of virtual reality can be connected directly with all possible commands and all types of translation. To demonstrate the possibilities of such a tool, a specific example of bridge was chosen: the bridge of Chevire (Nantes, France) which an overview is shown in Fig. A.1.

It is a typical example where the monitoring work of a bridge can be automated. This bridge has three distinct parts to watch for:

- On deck
- The bearings that support the deck: abutments at both ends and a dozen piles
- The foundation that allows the transmission of forces of structure and land.

Monitoring of the deck, abutments and piers lends itself well to automation by monitoring with an airborne platform. In the simulated mission, the airship should fly from its point of rest, watch the first bridge pier in translation, monitor the span and continue monitoring the second cell, eventually returning to another point of rest. To perform the simulation, the position in space (x, y, z) and orientation (Euler angles: roll, pitch, yaw) of the vehicle are sent to the interface located in a Matlab-Simulink @ model with an "s-function". Subsequently, the interface converts the data into the coordinate system of the viewer selected. This is a model which is used to calculate the data needed to display graphics. This model consists of several blocks from library sources. These blocks represent the mission planning, trajectory generation, the airship model and 3D visualization.

Fig. A.2 Whole trajectory without wind

Fig. A.3 Trajectory without wind with a different point of view

Several views are possible: on the road, on the airship, the whole trajectory and an distant observer. It is possible to move around the stage with arrows separated by 45 degrees: North, Northeast, East, South East, South, Southwest, West, North-West. Thus, it is possible to follow the airship in its motion. Figure A.2 shows the entire path followed by the autonomous airship in nominal conditions while Fig. A.3 presents the same trajectories with a different point of view.

Figure A.4 shows a simulation where the breeze is supposed to have a constant speed in directions x and z. It is easy to see that the influence of wind increases with time, because uncorrected errors are added to each other in following the initial flight plan.

The feasibility of the trajectory generated depends on the technique used, the cost function chosen and the various constraints. In this case, constraints on the airship concerning its geometry, kinematics and dynamics have to be considered. The constraints from the environment are mainly non-collision with fixed obstacles cumbersome environment and taking into account proximity interactions with the airship. If the obstacle avoidance depends on the geometry of the environment and is common to all robotic tasks, the second point depends in general physical characteristics of the airship. The criteria to be met during the troubleshooting planning

Fig. A.4 High velocity wind

concern that a solution must optimize a cost function expressed in terms of distance traveled by the airship between two configurations ends, duration or energy necessary to the execution of his movement. Other criteria may also be considered such as the inclusion of safety distance to obstacles.

The objectives of the planning function are:

- Ordering the passage on the various mission areas
- Calculate a path between each element of the route
- Require the implementation of the monitoring operation.

A.4 Monitoring of High Voltage Power Networks

This research project concerns the monitoring of overhead lines and very high voltage of 63 KV to 400 KV and dams and reservoirs that produce hydroelectric power. These construction works of civil engineering present a great diversity, both tied to their inherent construction, their geographical location and the purpose for which they are constructed. It is clear that the control and monitoring of these structures is essential to ensure sustainability so that they can meet the requirements for safe use. Today, the inspection is to diagnosing disorders that can present these works, to characterize them and follow them over time so as to respond in a timely manner and especially before the construction work reaches a critical state. Each year, in France, more than 100,000 km of lines are inspected as well. 400 KV lines and dams are under constant surveillance. The control principle is to be ready to intervene regardless of the date and time when the network situation requires it. Moreover, the resumption of activity after a climatic phenomenon must be done as quickly as possible while it is currently dangerous for inspectors. The inspection includes, therefore, two essential features

- Go to collect information on the ground by a systematic survey of all the disorders
- Process, analyze and leverage this information to know precisely the pathology and the exact condition of the structure, to use the appropriate time by the manager for the restoration.

In addition, inspection of these structures is a complex process of responding to strong obligations and standard requiring constant adaptation to the multiple types of electrical networks:

- Mesh (for the highest voltage)
- Radial or curly
- Tree...

To access all parts of the work, the inspection team makes use of heavy equipment resources (helicopters, boats...) sometimes endangering the lives of the officers conducting the inspection, not to mention the time inspection and the financial impact through the implementation of Security Officers in support of the actual inspection of the area around the structure, maintenance of devices. After an initial detailed inspection to classify the works, other detailed inspections are carried out with regular surveillance at intervals corresponding to the condition and age of the structure. In addition, there may be special inspection activities on the construction works whose condition is considered critical. Often, this means following specific areas of pathology to take the decision of repair or consolidation.

A.4.1 Current Market for Inspection of Electrical Networks

Once a year, technicians monitor the entire estate by helicopter. This procedure requires the presence of three officers on board a helicopter, a navigator, a pilot and a technical observer. The navigator advises the pilot on the route to follow. The pilot must fly close to lines, about fifteen or twenty meters above the ground, approximately 10 m above the electrical network. He must maneuver the helicopter to provide the best visibility to the technical observer. The technical agent visually detects anomalies and then photographed them for record and send response to the ground teams with overhanging structures, verify the anomalies on the lines and see if there are broken or cracked insulators by lightning, frayed conductors, tree branches on the lines or dangerously close to the lines, twisted wire broken, lightning exploded, transformer leaking... and must avoid the traps: the crossing of lines. ...This is a rigorous process and really stressful for the inspection team. In addition, electromagnetic fields generated by power lines, difficult weather (storm, snow...) and the danger of electrocution (electrical arcing can occur even without direct contact with the line) make this inspection operation also dangerous. This aerial inspection is done primarily for reasons of safety and security. It is accompanied by ground visits, inspection and control of towers by thermography. The latter technique to detect the temperature rise is also done by helicopter. The Technical Officer has a software coupled to a GPS to record the problems and keep the history. A good steering precision allows the observation of items to be inspected. The helicopter is currently instrumented as follows:

- The exterior: a sling designed to carry and stow items of equipment needed to repair lines and poles. A system of mirrors allows the pilot to control the maneuvers with precision

- The interior: a GPS screen that displays each component of the line (this system connected to a laptop allows visual reference) and an infrared camera that detects hot spots of the line, areas of excessive resistance to the passage current and likely to cause malfunctions.

Control and monitoring of high voltage lines are clearly a separate market, relatively large and yet this market remains poorly industrialized since only visual.

A.4.2 Project Goals

This research program focuses on the automation of inspection tasks, replacing the pilot and navigator in the inspection task and then providing specialized airborne robots for surveillance in general. This project is the integration of a different methodology from that currently used. The primary objective is to ensure the safety of users and distribution of electricity. Maintaining the infrastructure for maintenance as effective as possible is a key topic, then, a longevity/maintenance cost as high as possible.

Given the great diversity of structures, the human-made diagnosis has now reached certain limits. Indeed, an inspection requires a shift on the whole and then analyzing each of the previous and newly identified faults. This process requires expensive means of access, is long and lacks repeatability. This is especially relevant to tall towers. Requirements are:

- Statement of defects per unit length on the cables, or works hard to reach surfaces. The interest is to obtain a mapping of disorders. The need is to achieve a fast, economical seizure and treatment. Quantification need not be extremely precise.
- The Inspection Support: there is a strong interest in key areas to inspect for return simply an image that will be performed by the technical officer.
- Record malfunctioning geometric lines or towers.

The objective of this research project is to develop an aerial robotic diagnostic tool able to:

- Conduct an initial analysis which will determine whether human intervention is required (direct economic gain)
- Conduct sometimes this analysis in difficult wind conditions (security gain)
- To store the data acquired on the ground for further exploitation more relevant to the secure and reliable global diagnosis of the works.

This research offers a leap in both science and technology by offering a completely innovative technology based on airborne platform independent. One major innovation is the use of mini-UAV autonomous inspection of high voltage electrical networks, in difficult wind conditions, by establishing a process:

- Safer for those carrying out this diagnosis
- Simple to implement

- Less expensive than the current method
- More reliable diagnosis of structure.

Technological innovations will focus on the design of a mini autonomous UAV and its command and mission planning adapted to the inspection of high voltage networks. The system will have a remote control device in line with the work environment and a complete measurement, to assist the maintenance operator. A demonstrator must be proposed with a form suitable for inspection and payload sensors specific to this application. An innovative software to make the mission strategy by automating the inspection of high voltage in the presence of difficult flying conditions, should be proposed. The scientific and technical innovation will be presented through five main areas:

- Design of a lighter than air robot dedicated to this mission
- Autopilot
- Mission planning
- Gyro-stabilized turret to eliminate vibration and airborne vectors allow fine pointing toward the area to inspect.
- Image processing.

This research can be applied to other monitoring applications in the longer term.

A.5 FAA Recommendations

Tasks are knowledge areas, flight procedures or maneuvers appropriate to an area of operation

- Preflight procedures
  - Launch site selection
  - Inflation
  - Basket/Gondola management
  - Pre-launch check
- Launch and landing
  - Normal Launch
  - Launch over obstacle
  - Approach to landing
  - Normal landing
  - High-wind landing
- Fundamentals of flights
  - Altitude control (level flight)
  - Ascents
  - Descents
- Performance maneuvers
  - Rapid ascent
  - Rapid descent
  - Contour flying

  - High altitude flight
  - Obstruction clearance
- How to fly a balloon
  - Physical laws applicable to balloon flight
  - Effects of changes in temperature, pressure, humidity and altitude on maintaining equilibrium
  - Effects of false or uncontrolled lift during takeoff, landing and wind shear penetration
  - Fundamentals of a flight profile and course for intended route of flight based on winds aloft
- Mountain flying
  - Consideration for access to landing areas
  - Evidence of possible turbulence and descending air currents on leeward side of mountains
  - Concerns for terrain, effects on wind (upslope, downslope) and possible rapid weather changes
- Performance and limitations
  - Use of performance charts, tables and other data in determining performance in various phases of flight
  - Determine load and altitude limits
  - Requirement to arrest a terminal velocity descent
  - Effects of atmospheric conditions on performance
  - Factors to be considered in determining that the required performance is within the balloon's capabilities
- Launch site selection
  - Importance of size and surface condition of site
  - Consideration of accessibility and obstacles
  - Hazards surrounding launch site
  - Consideration of suitable landing areas based on wind conditions.

A.6 Indoor Lighter than Air Robot: A Differential Geometry Modeling Approach

If the assumption of filtering out the most of the pitch and roll motion, the state space of the lighter than air robot may be considered as $SE(2) \times \mathfrak{R}$, which is the product of the planar motion with a purely vertical translation, a proper subgroup of $SE(3)$. Let the transformation from the inertial fixed frame to the body fixed frame be

$$X_f = A^b_{M_f} X_b = \begin{pmatrix} R & \eta_1 \\ 0_{3\times 3} & 1 \end{pmatrix} X_b \tag{A.1}$$

where R is the rotation matrix and η_1 is the translation. The twist describing the velocity (or angular velocity) is denoted by ξ and $\hat{\xi} = Sk(\xi)$ matrix rotation such that

$$\hat{\xi} = Sk(\xi) = \begin{pmatrix} 0 & -\xi_3 & \xi_2 \\ \xi_3 & 0 & -\xi_1 \\ -\xi_2 & \xi_1 & 0 \end{pmatrix} \tag{A.2}$$

then the body velocity of the lighter than air robot is

$$\hat{\xi} = \left(A^b_{M_f}\right)^{-1} \dot{A}^b_{M_f} = \begin{pmatrix} \hat{\omega} & V \\ 0 & 0 \end{pmatrix} \tag{A.3}$$

where $V = (u, v, w)^T$ and $\omega = (p, q, r)$. Since there is no direct forcing in the side-slip direction, the external forces such as air current are neglected and since the large surface area provides a significant amount of drag, for an indoor lighter than air robot, it is considered that $p = q = v = 0$ [110, 226].

The dynamics of the indoor lighter than air robot is slow (≤ 3 m/s) and so the Reynolds number is relatively small ($\approx 10^4$). Thus it can be modeled as a neutrally buoyant rigid body in an ideal fluid. The body is approximated as an ellipsoid. Let P_f and Π_f denote respectively the linear and angular components of the momentum with respect to the inertial frame while P_b and Π_b are the same components written in the lighter than air frame, then

$$\begin{pmatrix} P_f \\ \Pi_f \end{pmatrix} = \begin{pmatrix} R & 0 \\ \hat{\eta}_1 R & R \end{pmatrix} \begin{pmatrix} P_b \\ \Pi_b \end{pmatrix} \tag{A.4}$$

or

$$\begin{aligned} P_f &= R P_b \\ \Pi_f &= R\Pi_b + \eta_1 \times P_f \end{aligned} \tag{A.5}$$

Let f_i, τ_j be the applied forces and torques in the fixed frame and F_i, Θ_j be those expressed in the lighter than air robot with $F_i = Rf_i$ and $\Theta_j = T\tau_j$. According to Newton's law,

$$\begin{aligned} \dot{P}_i &= \sum_i f_i \\ \dot{\Pi}_f &= \tau + R \sum_i f_i \times r_i \end{aligned} \tag{A.6}$$

where r_i is the vector from origin of the fixed frame to the line of action of the acting force and $\tau = \sum_i \tau_j$. The vector from the origin of the lighter than air frame to the same line is denoted with $\rho_i = r_i - \eta_1$. Differentiating, the following relations are obtained:

$$\begin{aligned} \dot{P}_b &= P_b \times \Omega_f + \sum_i f_i \\ \dot{\Pi}_b &= \tau + \Pi \times \omega + P_b \times V + \sum_i \rho_i \times F_i \end{aligned} \tag{A.7}$$

Next, the augmented mass and inertia matrices are respectively approximated by $M_a = \mathrm{diag}(m_x, m_y, m_z)$ and $J_a = (J_{ax}, J_{ay}, J_{az})$ where the cross-coupling terms between linear and angular velocity since the rotational motion is assumed to occur about a single axis. Then the reduced Lagrangian for the system is given by the kinetic energy of the lighter than air robot plus the air

$$\ell = \frac{1}{2}\left(\omega^T J\omega + V^T MV\right) \tag{A.8}$$

with

$$M = M_a + mI_{3\times 3}$$

and

$$J = J_a + J_{body}$$

Differentiating the Lagrangian with respect to V and Ω gives

$$P = \frac{\partial \ell}{\partial V} = MV$$

and

$$\Pi = \frac{\partial \ell}{\partial \omega} = J\omega$$

Noting that

$$F_1 = \begin{pmatrix} F_1 \cos\sigma \\ 0 \\ F_1 \sin\sigma \end{pmatrix}, \quad F_2 = \begin{pmatrix} F_2 \cos\sigma \\ 0 \\ F_2 \sin\sigma \end{pmatrix}, \quad F_3 = \begin{pmatrix} 0 \\ F_3 \\ 0 \end{pmatrix},$$

$$V = \begin{pmatrix} u \\ 0 \\ w \end{pmatrix}, \quad \omega = \begin{pmatrix} 0 \\ q \\ 0 \end{pmatrix}$$

Furthermore, the assumptions made above lead to

$$\begin{aligned} M_{11}\dot{u} &= (F_1 + F_2)\cos\sigma + F_{drag_x} \\ M_{33}\dot{w} &= (F_1 + F_2)\sin\sigma + F_{drag_z} \\ J_{22}\dot{q} &= (F_1 - F_2)\left(k_t l_x + l_y \cos\sigma\right) + \tau_{drag_z} \end{aligned} \tag{A.9}$$

where F_1, F_2 are actuating forces of the motors, $F_{drag_x}, F_{drag_z}, \tau_{drag_z}$ are drag forces due to air resistance, σ is the tilt angle of the propellers, l_x, l_y are acting lengths. Also the rotor forces has been included as $F_3 = k_t(F_1 - F_2)$ by design where k_t is predetermined.

The dynamic equations of the system may be written as:

$$\xi = \begin{pmatrix} \dot{u} \\ \dot{w} \\ \dot{q} \end{pmatrix}$$

$$
= \begin{pmatrix} M_{11} & 0 & 0 \\ 0 & M_{33} & 0 \\ 0 & 0 & J_{22} \end{pmatrix}^{-1} \begin{pmatrix} B_1 & 0 & 0 \\ 0 & B_2 & 0 \\ 0 & 0 & B_3 \end{pmatrix} \xi^2
$$

$$
+ \begin{pmatrix} M_{11} & 0 & 0 \\ 0 & M_{33} & 0 \\ 0 & 0 & J_{22} \end{pmatrix}^{-1} \begin{pmatrix} (F_1 + F_2)\cos\sigma \\ (F_1 + F_2)\sin\sigma \\ (F_1 - F_2)\left(k_t l_x + l_y \cos\sigma\right) \end{pmatrix} \quad \text{(A.10)}
$$

where

$$
\begin{aligned}
F_{drag_x} &= B_1 u^2 \\
F_{drag_z} &= B_2 w^2 \\
\tau_{drag_x} &= B_3 q^2
\end{aligned}
$$

References

1. Adamski, W., Herman, P., Bestaoui, Y., Kozlowski, K.: Control of an airship in case of unpredictable environment conditions. In: IEEE Conf. on Control and Fault-Tolerant Systems, Nice, France, pp. 843–848 (2010)
2. AIAA-G003C-2010 Guide to reference and standard atmosphere models (2010)
3. Alighanbari, M., Bertuccelli, L.F., How, J.P.: A robust approach to the UAV task assignment problem. In: IEEE Conf. on Decision and Control, San Diego, CA, pp. 5935–5940 (2006)
4. Ambrosino, G., Ariola, M., Ciniglio, U., Carraro, F., Delellis, E., Pironti, A.: Path generation and tracking in 3D for UAVs. IEEE Trans. Control Syst. Technol. **17**, 980–988 (2009)
5. Ardema, M.D.: Missions and vehicle concepts for modern, propelled lighter than air vehicles. NASA Technical Memo 86672, Ames Research Center, CA (1984)
6. Avanzini, G.: Frenet based algorithm for trajectory prediction. J. Guid. Control Dyn. **27**, 127–135 (2004)
7. Azinheira, J.R., Moutinho, A.B.: Hover control of an UAV with backstepping design including input saturations. IEEE Trans. Control Syst. Technol. **16**, 517–526 (2007)
8. Azinheira, J.R., de Paiva, E.C., Bueno, S.S.: Influence of wind speed on airship dynamics. J. Guid. Control Dyn. **25**, 1116–1124 (2002)
9. Azinheira, J.R., Moutinho, A.B., de Paiva, E.C.: A backstepping controller for path tracking of an underactuated autonomous airship. Int. J. Robust Nonlinear Control **19**, 418–441 (2009)
10. Azouz, N., Bestaoui, Y., Lemaitre, O.: Dynamic analysis of airships with small deformations. In: 3rd IEEE Workshop on Robot Motion and Control, Bukowy-Dworek, pp. 209–215 (2002)
11. Bairstow, L.: Applied Aerodynamics. Longmans, Green, London (1920)
12. Bateman, H.: The inertia coefficients of an airship in a frictionless fluid. Technical report 164, NACA (1924)
13. Belta, C., Kumar, V.: An SVD based projection method for interpolation on SE(3). IEEE Trans. Robot. Autom. **18**, 334–345 (2002)
14. Bestaoui, Y.: On line reference trajectory definition with joint torque and velocity constraints. Int. J. Robot. Res. **11**, 75–85 (1992)
15. Bestaoui, Y.: Modeling and control of small autonomous airships. In: Castillo, P., Lozano, R., Dzul, A. (eds.) Modeling and Control of Mini Flying Machines. Springer, Berlin (2005)
16. Bestaoui, Y.: Nominal trajectories of an autonomous under-actuated airship. Int. J. Control. Autom. Syst. **4**, 395–404 (2006)
17. Bestaoui, Y.: Modeling and trajectory generation of lighter than air aerial robot. In: Kozlowski, K. (ed.) Robot Motion and Control 2007. Lectures Notes in Control and Information Sciences, vol. 360, pp. 3–28. Springer, Berlin (2007)

Y. Bestaoui Sebbane, *Lighter than Air Robots*,
Intelligent Systems, Control and Automation: Science and Engineering 58,
DOI 10.1007/978-94-007-2663-5, © Springer Science+Business Media B.V. 2012

18. Bestaoui, Y.: Geometric Properties of Aircraft Equilibrium and Non Equilibrium Trajectory Arcs. Kozlowski, K. (ed.) Lectures Notes in Control and Information Sciences. Springer, Berlin (2009)
19. Bestaoui, Y.: Lighter than air vehicle: autonomous airships for bridge monitoring. In: Columbus, F. (ed.) Helium: Characteristics, Compounds and Applications. Nova Publ., New York (2010)
20. Bestaoui, Y.: Mission plan under uncertainty for an autonomous aircraft. Proc. Inst. Mech. Eng., G J. Aerosp. Eng. **24**, 1297–1307 (2010)
21. Bestaoui, Y.: Bridge monitoring by a lighter than air robot. In: AIAA Aerospace Sciences Meeting, Orlando, FL (2011)
22. Bestaoui, Y., Dicheva, S.: 3D flight plan for an autonomous aircraft. In: 48th AIAA Aerospace Sciences Meeting, Orlando, FL, Paper AIAA-1352 (2010)
23. Bestaoui, Y., Hamel, T.: Dynamic modeling of small autonomous blimps. In: Conference Methods and Models in Automation and Robotics, Miedzyzdroje, Poland, pp. 579–584 (2000)
24. Bestaoui, Y., Hima, S.: Some insights in path planning of small autonomous blimps. Arch. Control Sci. **11**, 139–166 (2001)
25. Bestaoui, Y., Kahale, E.: Analysis of time optimal 3D paths for an autonomous aircraft with a piecewise constant acceleration. In: AIAA Aerospace Sciences Meeting, Orlando, FL (2011)
26. Bestaoui, Y., Kuhlmann, K.: Modeling of a quad-rotor airship with wind and freight effect: a Newton Euler approach. In: AIAA Aerospace Sciences Meeting, Orlando, FL (2010)
27. Bestaoui, Y., Lakhlef, F.: Flight plan for an autonomous aircraft in a windy environment. In: Lozano, R. (ed.) Unmanned Aerial Vehicles Embedded Control. Wiley, New York (2010)
28. Bestaoui, Y., Hima, S., Sentouh, C.: Motion planning of a fully actuated unmanned air vehicle. In: AIAA Conference on Navigation, Guidance and Control, Austin, TX (2003)
29. Bestaoui, Y., Dahmani, H., Belharet, K.: Geometry of translational trajectories for an autonomous aerospace vehicle with wind effect. In: 47th AIAA Aerospace Sciences Meeting, Orlando, FL, Paper AIAA-1352 (2009)
30. Bethke, B., Valenti, M., How, J.P.: UAV task assignment, an experimental demonstration with integrated health monitoring. IEEE J. Robot. Autom. **15**, 39–44 (2008)
31. Bicho, E., Moreira, A., Carvalheira, M., Erlhagen, W.: Autonomous flight trajectory generator via attractor dynamics. In: Proc. of IEEE/RSJ Intelligents Robots and Systems, vol. 2, pp. 1379–1385 (2005)
32. Blackmore, L., Ono, M., Bektassov, A., Williams, B.: A probabilistic particle control approximation of chance constrained stochastic predictive control. IEEE Trans. Robot. **26**, 502–517 (2010)
33. Blanchini, F., Sznaier, M.: Rational L1 suboptimal compensators for continuous time systems. IEEE Trans. Autom. Control **39**, 1487–1492 (1994)
34. Bloch, A.M.: Non Holonomics Mechanics and Control. Springer, Berlin (2003)
35. Bollino, K.P., Lewis, L.R., Sekhavat, P., Ross, I.M.: Pseudo-spectral optimal control: a clear road for autonomous intelligent path planning. In: AIAA Infotech@aerospace Conference, Rohnert Park, CA, AIAA 2007-2831 (2007)
36. Bonnet, A.: Identification des coefficients aerodynamiques du dirigeable AS500 du LAAS. Technical report, Etude Hydro-Aerodynamique, SUPAERO (2003)
37. Boukraa, D., Bestaoui, Y., Azouz, N.: 3D dimensional trajectory tracking for a fixed wing unmanned aerial vehicle using dynamic inversion. In: 17th IFAC Symposium on Automatic Control in Aerospace, Toulouse, France (2007)
38. Boukraa, D., Bestaoui, Y., Azouz, N.: Three dimensional trajectory generation for an autonomous plane. Inter. Review of Aerospace Eng. **4**, 355–365 (2008)
39. Brockett, R.W.: Asymptotic stability and feedback stabilization. In: Brockett, R.W., Millman, R.S., Sussmann, H.J. (eds.) Differential Geometric Control Theory, pp. 181–191. Birkhäuser, Basel (1983)
40. Bueno, S.S., dePaiva, E.C., Azinheira, J.R., Ramos, J.J.G., Carvalho, J.R.H., Elfes, A., Rives, P., Silveira, G.S.: Project AURORA—towards an autonomous robotic airship. In: IEEE/RSJ Int. Conf. on Intelligent Robots and Systems, Lausanne, Switzerland (2002)

41. Burgess, C.P.: Forces on airships in gusts. Report N. 24, Bureau of Aeronautics, Navy Dept. (1934)
42. Cai, Z., Qi, W., Xi, Y.: Dynamic modeling for airship equipped with ballonets and ballasts. Appl. Math. Mech. **26**, 1072–1082 (2005)
43. Canudas de Wit, C., Siciliano, B., Bastin, G. (eds.): Theory of Robot Control. Springer, New York (1996)
44. Chen, B., Nagarajaiah, S.: Linear matrix inequality based robust fault detection and isolation using the eigenstructure assignment method. J. Guid. Control Dyn. **30**, 1831–1835 (2007)
45. Choset, H., Lynch, K., Hutchinson, S., Kantor, G., Burgard, W., Kavraki, L., Thrum, S.: Principles of Robot Motion, Theory, Algorithms and Implementation. MIT Press, Cambridge (2005)
46. Colgren, R.D.: Applications of Robuts Control to Nonlinear Systems. Progress in Astronautics and Aeronautics. AIAA, Washington (2004)
47. Conte, G., Moog, C.H., Perdon, A.M.: Algebraic Methods for Nonlinear Control Systems: Theory and Applications. Springer, New York (2006)
48. Cook, M.V., Lipscombe, J.M., Goineau, F.: Analysis of the stability modes of the non rigid airship. Aeronaut. J. **104**, 279–289 (2000)
49. Coron, J.-M.: On the stabilization of some nonlinear control systems: results, tools, and applications. NATO Advanced Study Institute, Montreal (1998)
50. Crump, M.R.: The dynamics and control of catapult launching unmanned air vehicles from moving platforms. Ph.D. thesis, RMIT University, Melbourne, Australia (2002)
51. D'angelo, S., Minisci, E., DiBona, D.: Optimization methodology for ascent trajectories of lifting body reusable launchers. J. Spacecr. Rockets **37**(6), 761–767 (2000)
52. Das, T., Mukherjee, R., Cameron, J.: Optimal trajectory planning for hot air balloons in linear wind fields. J. Guid. Control Dyn. **26**(3), 416–424 (2003)
53. Davis, J.D., Chakravorty, S.: Motion planning under uncertainty: application to an unmanned helicopter. J. Guid. Control Dyn. **30**, 1268–1276 (2007)
54. Dicheva, S., Bestaoui, Y.: Route finding for an autonomous aircraft. In: AIAA Aerospace Sciences Meeting, Orlando, FL (2011)
55. Donneker, S.: Distributed thrust and autonomous ground handling. In: 13th AIAA Lighter than Air Systems Technology Conference, pp. 122–131 (1999)
56. Doyle, J.C., Stein, G.: Multivariable feedback design: concepts for a classical/modern synthesis. IEEE Trans. Autom. Control **26**, 4–16 (1981)
57. Doyle, J.C., Glover, K., Khargonekar, P.P., Francis, B.A.: State-space solutions to standard H2 and H_∞ control problems. IEEE Trans. Autom. Control **34**, 831–847 (1989)
58. Dubins, L.E.: On curves of minimal length with a constraint on average curvature and with prescribed initial and terminal positions and tangents. Am. J. Math. **79**, 497–517 (1957)
59. Eele, A., Richards, A.: Path planning with avoidance using nonlinear branch and bound optimization. J. Guid. Control Dyn. **32**, 384–394 (2009)
60. Elfes, A., Bueno, S.S., Bergerman, M., Ramos, J.J.G.: A semi-autonomous robotic airship for environmental monitoring missions. In: Inter. Conf. on Robotic and Automation, Leuven, Belgium (1998)
61. Elfes, A., Bueno, S.S., Bergerman, M., Ramos, J.J.G.: Robotic airship for exploration of planetary bodies with an atmosphere: autonomy challenges. Auton. Robots **14**, 147–164 (2003)
62. Elfes, A., Hall, J.L., Kulczycki, A.: Autonomy architecture for aerobot exploration of Saturnian moon Titan. IEEE Aerosp. Electron. Syst. Mag. **23**, 16–24 (2008)
63. Etkin, B.: Dynamics of Atmospheric Flight. Dover, New York (2000)
64. Evans, J.R., DeLaurier, J.D., Scholaert, H.: A study of airship six degrees of freedom flight dynamics. Research report N. 79, University of Toronto, Canada (1981)
65. Fahimi, F.: Autonomous Robots: Modeling, Path Planning and Control. Springer, New York (2009)
66. Fantoni, I., Lozano, R.: Nonlinear Control for Under-actuated Mechanical Systems. Springer, New York (2002)
67. Farault, J.: Analysis on Lie Groups: An Introduction. Cambridge Studies in Advanced Mathematics (2008)

68. Farouki, R.T., Giannelli, C., Manni, C., Sestini, A.: Identification of spatial PH quintic Hermite interpolants with near-optimal shape measures. Comput. Aided Geom. Des. **25**, 274–297 (2008)
69. Fraichard, T., Scheuer, A.: From Reeds and Shepp's to continuous curvature paths. IEEE Trans. Robot. **20**, 1025–1035 (2008)
70. Franklin, G.F., Powell, J.D., Workman, M.: Digital Control of Dynamic Systems. Addison-Wesley, Reading (1998)
71. Frazzoli, E.: Robust hybrid control for autonomous vehicle motion planning. Ph.D. thesis, MIT, Cambridge, MA (2001)
72. Frazzoli, E., Dahleh, M.A., Feron, E.: Maneuver based motion planning for nonlinear systems with symmetries. IEEE Trans. Robot. **4**, 355–365 (2008)
73. Fossen, T.I.: Guidance and Control of Ocean Vehicle. Wiley, New York (1996)
74. Fossen, T.I., Ross, A.: Nonlinear modeling, identification and control of UUVs. In: Roberts, G., Sutton, B. (eds.) Guidance and Control of Unmanned Marine Vehicles. IEE Control Engineering Series, pp. 23–42 (2006)
75. Fukao, T., Kanzawa, T., Osuka, K.: Inverse optimal tracking control of an aerial blimp robot. In: IEEE Workshop on Robot Motion and Control, pp. 193–198 (2005)
76. Fukushima, H., Saito, R., Matsuno, R., Hada, Y., Kawabata, K., Asama, H.: Model predictive control of an autonomous blimp with input and output constraints. In: IEEE Int. Conf. on Control Applications, Munich, Germany, pp. 2184–2189 (2006)
77. Fukushima, H., Kon, K., Hada, Y., Matsuno, R., Kawabata, K., Asama, H.: State predictive control of an autonomous blimp in the presence of time delay and disturbance. In: IEEE Int. Conf. on Control Applications, pp. 188–193 (2007)
78. Goerzen, C., Kong, Z., Mettler, B.: A survey of motion planning algorithms from the perspective of autonomous UAV guidance. J. Intell. Robot. Syst. **20**, 65–100 (2010)
79. Gomes, S.B.V.: Recent airship technology developments in Brazil. Airship **167**, 21–24 (2010)
80. Gomes, S.B.V., Ramos, J.G.: Airship dynamic modeling for autonomous operation. In: IEEE Int. Conf. on Robotics and Automation, vol. 4, pp. 3362–3467 (1998)
81. Greengard, G., Ruszczynski, A.: Decision Making Under Uncertainty: Energy and Power. Springer, Berlin (2000)
82. Hima, S.: Planification de trajectories d'un dirigeable autonome. Ph.D. thesis, Univ. of Evry, France (Dec. 2005)
83. Hima, S., Bestaoui, Y.: Motion generation on trim trajectories for an autonomous underactuated airship. In: 4th Inter. Airship Conf., Cambridge, England, July 2002
84. Hima, S., Bestaoui, Y.: Time optimal paths for lateral navigation of an autonomous underactuated airship. In: AIAA Conference on Navigation, Guidance and Control, Austin, TX, Aug. 2003
85. Hima, S., Bestaoui, Y.: Trim trajectories characterization for an unmanned autonomous airship. In: IEEE/RSJ Int. Conf. on Intelligent Robots and Systems IROS 2006, Taiwan
86. Holdsworth, R.: Autonomous in flight path planning to replace pure collision avoidance for free flight aircraft using automatic dependent surveillance broadcast. Ph.D. thesis, Swinburne Univ. (2003)
87. Hong, C.H., Choi, K.C., Kim, B.S.: Applications of adaptive neural network control to an unmanned airship. Int. J. Control. Autom. Syst. **7**, 911–917 (2009)
88. Hyde, R.: The application of robust control to VSTOL aircraft. Ph.D. thesis, Cambridge Univer., United Kingdom (1991)
89. Hygounenc, E.: Modelisation et Commande d'un Dirigeable pour le Vol Autonome. Ph.D. thesis, LAAS-CNRS, France (2003)
90. Hygounenc, E., Soueres, P.: Lateral path following GPS-based control of a small size unmanned blimp. In: IEEE Int. Conf. on Robotics and Automation, pp. 540–545 (2002)
91. Jaklic, G., Kozak, J., Krajnc, M., Vitrih, V., Zagar, E.: Geometric Lagrange interpolation by planar cubic Pythagorean hodograph curves. Comput. Aided Geom. Des. **25**, 720–728 (2008)

92. Jardin, M.R., Bryson, A.E.: Neighboring optimal aircraft guidance in winds. J. Guid. Control Dyn. **24**, 710–715 (2001)
93. Jennings, A.L., Ordonez, R., Ceccarelli, N.: Dynamic programming applied to UAV way point path planning in wind. In: IEEE International Symposium on Computer-Aided Control System Design, San Antonio, TX, pp. 215–220 (2008)
94. Jia, R., Frye, M.T., Qian, C.: Control of an airship using particle swarm optimization and neural network. In: IEEE Inter. Conf. on Systems, Man and Cybernetics, San Antonio, TX, pp. 1809–1814 (2009)
95. Jiang, Z., Ordonez, R.: Robust approach and landing trajectory generation for reusable launch vehicles in winds. In: 17th IEEE International Conference on Control Applications, San Antonio, TX, pp. 930–935 (2008)
96. Jurdjevic, V.: Geometric Control Theory. Cambridge Studies in Advanced Mathematics (2008)
97. Kala, R., Shukla, A., Tiwari, R.: Fusion of probabilistic A* algorithm and fuzzy inference system for robotic path planning. Artif. Intell. Rev. **33**, 307–327 (2010)
98. Kaminer, I., Khargonekar, P.P., Robel, G.: Design of a localizer capture and track modes for a lateral autopilot using H_∞ synthesis. IEEE Control Syst. Mag. **10**, 13–21 (1990)
99. Kaminer, I., Pascoal, A., Hallberg, E., Silvestre, C.: Trajectory tracking for autonomous vehicles: an integrated approach to guidance and control. J. Guid. Control Dyn. **21**, 29–38 (1998)
100. Kampke, T., Elfes, A.: Optimal aerobot trajectory planning for wind based opportunity flight control. In: IEEE/RSJ Inter. Conf. on Intelligent Robots and Systems, Las Vegas, NV, pp. 67–74 (2003)
101. Kanayama, Y., Miyake, N.: Trajectory generation for mobile robots. In: Robotics Research, vol. 3, pp. 333–340 (1986)
102. Kantor, G., Wettergreen, D., Ostrowski, J.P., Singh, S.: Collection of environmental data from an airship platform. In: SPIE Conf. on Sensor Fusion and Decentralized Control in Robotic Systems IV, SPIE, Bellingham, WA, vol. 4571, pp. 76–83 (2001)
103. Kavraki, L., Latombe, J.C.: Randomized preprocessing of configuration space for fast path planning. In: IEEE Inter. Conf. on Robotic and Automation, vol. 3, pp. 2138–2145 (1994)
104. Kawano, H.: Three dimensional obstacle avoidance of autonomous blimp flying in unknown disturbance. In: IEEE/RSJ Inter. Conf. on Intelligent Robots and Systems, pp. 123–130 (2006)
105. Kayuk, Y., Denisenko, V.: Motion of a mechanical system with variable mass inertia characteristics. Inter. J. Appl. Mech. **40**, 814–820 (2004)
106. Khammash, M., Zou, L., Almquist, J.A., Van Der Linden, C.: Robust aircraft pitch axis control under weight and center of gravity uncertainty. In: 38th IEEE Conference on Decision and Control, vol. 2, pp. 1970–1975 (1999)
107. Khatib, O.: Real time obstacle avoidance for manipulators and mobile robots. In: IEEE Inter. Conf. on Robotics and Automation (1985)
108. Khoury, G., Gillett, J.D.: Airship Technology. Cambridge Aerospace Series (1999)
109. Kim, J., Khosla, P.K.: Real time obstacle avoidance using harmonic potential functions. IEEE Trans. Robot. Autom. **8**, 338–349 (1992)
110. Kim, J., Ostrowski, J.P.: Motion planning of an aerial robot using rapidly exploring random trees with dynamic constraints. In: IEEE Inter. Conf. on Robotics and Automation, vol. 2, pp. 2200–2205 (2003)
111. Kim, J., Keller, J., Kumar, V.: Design and verification of controllers for airships. In: IEEE/RSJ Inter. Conf. on Intelligent Robots and Systems, Las Vegas, NV, pp. 54–60 (2003)
112. Ko, J., Klein, D.J., Fox, D., Haehnel, D.: Gaussian process and reinforcement learning for identification and control of an autonomous blimp. In: Int. Conf. on Robotics and Automation, pp. 742–747 (2007)
113. Kornienko, A., Well, K.: Estimation of longitudinal motion of a remotely controlled airship. In: AIAA Atmospheric Flight Mechanics Conf., Austin, TX (2003)
114. Kozlowski, K. (ed.): Robot Motion and Control: Recent Developments. Springer, Berlin (2006)

115. Krozel, J., Penny, S., Prete, J., Mitchell, J.S.B.: Automated route generation for avoiding deterministic weather in transition airspace. J. Guid. Control Dyn. **30**, 144–153 (2007)
116. Kungl, P., Schlenker, M., Winner, D.A., Kroplin, B.H.: Instrumentation of remote controlled airship Lotte for in-flight measurements. Aerosp. Sci. Technol. **8**, 599–610 (2004)
117. Kurtoglu, T., Johnson, S.B., Barszcz, E., Johnson, J.R., Robinson, P.I.: Integrating system health management into the early design of aerospace system using functional fault analysis. In: Inter. Conf. on Prognostic and Health Management (2008)
118. Kuwata, Y., Blackmore, L., Wolf, M., Fathpour, N., Newman, C., Elfes, A.: Decomposition algorithm for global reachability analysis on a time varying graph with an application to planetary exploration. In: IEEE/RSJ Int. Conf. on Intelligent Robots, pp. 3955–3960 (2009)
119. Kwakernaak, H.: Robust control and H_∞ optimization—tutorial paper. Automatica **29**, 255–273 (1993)
120. Lacroix, S., Jung, I.K.: High resolution terrain mapping with an autonomous blimp. In: IEEE/RSJ Conf. on Intelligent Robots and Systems, vol. 1, pp. 781–786 (2002)
121. Lafferiere, G., Sussmann, J.H.: A Differential Geometric Approach to Motion Planning. Kluwer Academic, Dordrecht (1993)
122. Lam, T.M. (ed.): Aerial Vehicles. In-Tech, Vienna (2009)
123. Lamb, H.: On the Motion of Solids Through a Liquid. Hydrodynamics. Dover, New York (1945)
124. Laugier, C., Chatila, R. (eds.): Autonomous Navigation in Dynamic Environment. Springer, Berlin (2007)
125. Lavalle, S.M.: Planning Algorithms. Cambridge University Press, Cambridge (2006)
126. Lee, S., Bang, H.: 3D ascent trajectory optimization for stratospheric airship platforms in the jet stream. J. Guid. Control Dyn. **30**, 1341–1352 (2007)
127. Lee, Y.J., Kim, D.M., Yeom, C.H.: Development of Korean High altitude platform systems. Int. J. Wirel. Inf. Netw. **13**, 31–42 (2006)
128. Lennon, J.A., Atkins, E.M.: Multi-objective spacecraft trajectory optimization with synthetic agent oversight. J. Aerosp. Comput. Inf. Commun. **2**, 4–24 (2005)
129. Lentilhac, S.: UAV flight plan optimized for sensor requirements. IEEE Aerosp. Electron. Syst. Mag. **25**, 11–14 (2010)
130. Li, T., Zhang, Y., Zhang, Z.: Approaches to motion planning for a spherical robot based on differential geometric control theory. Paper presented at the 6th World Congress on Intelligent Control and Autonomation, Dalian, China, June 2006
131. Li, Y., Nahon, M.: Modeling and simulation of airship dynamics. J. Guid. Control Dyn. **30**, 1691–1700 (2007)
132. Li, Z., Canny, J.F.: Non Holonomic Motion Planning. Kluwer Academic, Berlin (1992)
133. Lia, L., Pasternak, I.: A review of airship structural research and development. Prog. Aerosp. Sci. (2009). doi:10.1016/j.paerosci.2009.03.001
134. Liu, Y., Pan, Z., Stirling, D., Naghdy, F.: Control of autonomous airship. In: IEEE Int. Conf. on Robotics and Biomimetics, Guilin, China, pp. 2457–2462 (2009)
135. Lorenz, R.D.: Flight power scaling of airplanes, airships and helicopters: application to planetary exploration. J. Aircr. **38**, 208–214 (2001)
136. Ludington, B., Johnson, E., Vachtsevanos, A.: Augmenting UAV autonomy (GTMAX). IEEE Robot. Autom. Mag. **21**, 67–71 (2006)
137. Luffman, C.R.: A dissertation on how non-rigid airships ascend and descend. Airship **164**, 21–28 (2009)
138. Lukbowski, S., Jones, B., Rojas, E., Morris, D.: Trade-off analysis of regenerative power source for long duration loitering airship. In: Systems and Information Engineering Design Symposium, Charlottesville, VA, pp. 25–30 (2010)
139. Lutz, T., Wagner, S.: Drag reduction and shape optimization of airship bodies. J. Aircr. **35**, 345–351 (1998)
140. Macharet, D., Neto, A.A., Campos, M.: On the generation of feasible paths for aerial robots in environments with obstacles. In: IEEE/RSJ Int. Conf. on Intelligent Robots and Systems, pp. 3380–3385 (2009)

141. Mantegh, I., Jenkin, M., Goldenberg, A.: Path planning for autonomous mobile robots using the boundary integral equation method. J. Intell. Robot. Syst. doi:10.1007/s10846-010-9394-y
142. Marigo, A., Bichi, A.: Steering driftless nonholonomic systems by control quanta. In: IEEE Inter. Conf. on Decision and Control, vol. 4, pp. 466–478 (1998)
143. Marsden, J., Ratiu, T.S.: Introduction to Mechanics and Symmetry. Springer, Berlin (1999)
144. McGee, T., Hedrick, J.K.: Optimal path planning with a kinematic airplane model. J. Guid. Control Dyn. **30**, 629–633 (2007)
145. McManus, J., Walker, R.: Multidisciplinary approach to intelligent unmanned airborne vehicles mission planning. J. Aircr. **43**(2), 318–335 (2006)
146. Meskin, N., Jiang, T., Sobhani, E., Khorasani, K., Rabbath, C.A.: A nonlinear geometric fault detection and isolation approach for almost lighter than air vehicles. In: IEEE Int. Conf. on Control Applications, pp. 1073–1078 (2007)
147. Miele, A., Wang, T., Melvin, W.: Optimal take-off trajectories in the presence of windshear. J. Optim. Theory Appl. **49**, 1–45 (1986)
148. Miele, A., Wang, T., Melvin, W.: Penetration landing guidance trajectories in the presence of windshear. J. Guid. Control Dyn. **12**, 806–814 (1989)
149. Milne, L.M.: Theoretical Aerodynamics. Dover, New York (1973)
150. Moutinho, A.B.: Modeling and nonlinear control for airship autonomous flight. Ph.D. thesis, Univ. Tech. Lisbon, Portugal (2007)
151. Moutinho, A.B., Azinheira, J.R.: Stability and robustness analysis of the AURORA airship control system using dynamic inversion. In: IEEE Int. Conf. on Robotics and Automation, vol. 3, pp. 2265–2270 (2005)
152. Mueller, J.B., Zhao, Y.J.: Development of an aerodynamic model and control law design for a high altitude airship. In: AIAA 3rd Unmanned Unlimited Technical Conference, Workshop and Exhibit, Chicago, IL (2004)
153. Mueller, J.B., Zhao, Y.J., Garrard, W.L.: Optimal ascent trajectories for stratospheric airships using wind energy. J. Guid. Control Dyn. **32**, 1232–1245 (2004)
154. Munk, M.: The aerodynamic forces on airship hulls. Report N. 184 NACA (1924)
155. Murray, R.M., Sastry, S.S.: Non-holonomic motion planning: steering using sinusoids. IEEE Trans. Autom. Control **38**, 700–716 (1993)
156. Murray, R.M., Li, Z., Sastry, S.S.: Mathematical Introduction to Robotic Manipulation. CRC Press, Boca Raton (1994)
157. Nagabushan, B., Tomlinson, N.: Thrust-vectored takeoff, landing and ground handling of an airship. J. Aircr. **23**, 250–254 (1986)
158. Nelson, R., Barber, B., McLain, T., Beard, R.: Vector field path following for miniature air vehicle. IEEE Trans. Robot. **23**, 519–529 (2007)
159. Nicolos, I.K., Valavanis, K.P., Tsourveloudis, N.T., Kostaras, A.N.: Evolutionary algorithm based offline/online path planner for UAV navigation. IEEE Trans. Syst. Man Cybern. **33**, 898–912 (2003)
160. Olfati-Saber, R.: Nonlinear control of under-actuated mechanical systems with application to robotics and aerospace vehicles. Ph.D. thesis, MIT, Cambridge, MA (2001)
161. Ollero, A.: Control and perception techniques for aerial robotics. In: 7th IFAC Symposium on Robot Control, Wroclaw, Poland (2003)
162. Ollero, A., Maza, I.: Multiple Heterogeneous Unmanned Aerial Vehicle. Springer, Berlin (2007)
163. Onda, M.: Design considerations on stratospheric LTA platform. In: 13th AIAA Lighter than Air Systems Technology Conference, pp. 204–209 (1999)
164. Park, C.S., Lee, H., Tahk, M.J., Bang, H.: Airship control using neural network augmented model inversion. In: IEEE Int. Conf. on Control Applications, vol. 1, pp. 558–563 (2003)
165. Pashilkar, A.A., Sundararajan, N., Saratchandran, P.: A fault-tolerant neural aided controller for aircraft auto-landing. Aerosp. Sci. Technol. **10**, 49–61 (2006)
166. Patsko, V.S., Botkin, N.D., Kein, V.M., Turova, V.L., Zarkh, M.A.: Control of an aircraft landing in windshear. J. Optim. Theory Appl. **83**, 237–267 (1994)

167. Pettersen, K.Y., Egeland, O.: Time-varying exponential stabilization of the position and attitude of an under actuated autonomous underwater vehicle. IEEE Trans. Autom. Control **44**, 112–115 (1999)
168. Pettersson, P.O., Doherty, P.: Probabilistic road map based path planning for an autonomous unmanned aerial vehicle. In: Workshop on Connecting Planning Theory with Practice (2004)
169. Phillips, J.M., Bedrossian, N., Kavraki, L.E.: Guided expansive space trees: a search strategy for motion and cost constrained state spaces. In: IEEE Int. Conf on Robotics and Automation, vol. 5, pp. 3968–3973 (2004)
170. Piazzi, A., Guarino Lo Bianco, C., Romano, M.: η^3 splines for the smooth path generation of wheeled mobile robot. IEEE Trans. Robot. **5**, 1089–1095 (2007)
171. Postlethwaite, I., Bates, D.: Robust integrated flight and propulsion controller for the Harriet aircraft. J. Guid. Control Dyn. **22**, 286–290 (1999)
172. Prasanth, R.K., Boskovic, J.D., Li, S.M., Mehra, R.: Initial study of autonomous trajectory generation for UAV. In: IEEE Inter. Conf. on Decision and Control, Orlando, FL, pp. 640–645 (2001)
173. Prentice, B.E., Matthew, M. (co-eds.): North American transportation networks: gaps and opportunities. In: Proceedings of the 42nd Annual Meetings of the Canadian Transportation Research Forum (2007)
174. Rabier, P.J., Rheinboldt, W.C.: Nonholonomic Motion of Rigid Mechanical Systems from a DAE Viewpoint. SIAM, Philadelphia (2000)
175. Ramos, J., Maeta, S., Mirisola, L., Bueno, S., Bergerman, M., Faria, B., Pinto, G., Bruciapaglia, A.: Internet based solutions in the development and operation of an unmanned robotic airship. Proc. IEEE **9**, 463–474 (2003)
176. Rao, J., Gong, Z., Luo, J., Xie, S.: A flight control and navigation system of a small size unmanned airship. In: IEEE Int. Conf. on Mechatronics and Automation, Niagara Falls, Canada, pp. 1491–1496 (2005)
177. Raymer, D.P.: Aircraft Design: A Conceptual Approach. AIAA Education Series (2006)
178. Richards, A., Schouwenaars, T., How, J., Feron, E.: Spacecraft trajectory planning with avoidance constraints using mixed-integer linear programming. J. Guid. Control Dyn. **25**, 755–764 (2002)
179. Rottmann, A., Plagemann, C., Hilgero, P., Burgaud, W.: Autonomous blimp control using model free reinforcement learning in a continuous state and action space. In: IEEE IROS, San Diego, CA, pp. 1895–1900 (2007)
180. Rysdyk, R.: Course and heading changes in significant wind. J. Guid. Control Dyn. **30**, 1168–1171 (2007)
181. Sastry, S.: Nonlinear Systems, Analysis, Stability and Control. Springer, Berlin (1999)
182. Schmidt, D.K.: Modeling and near-space station-keeping in control of a large high altitude airship. J. Guid. Control Dyn. **30**, 540–547 (2007)
183. Schouwenaars, T.: Safe trajectory planning of autonomous vehicles. Ph.D. thesis, MIT (2006)
184. Schouwenaars, T., Mettler, B., Feron, E.: Hybrid model for trajectory planning of agile autonomous vehicles. J. Aerosp. Comput. Inf. Commun. **12**, 466–478 (2004)
185. Seibel, C.W., Farines, J.M., Cury, J.E.: Towards hybrid automata for the mission planning of Unmanned Aerial Vehicles. In: Antsaklis, P.J. (ed.) Hybrid Systems V, pp. 324–340. Springer, Berlin (1999)
186. Selig, J.M.: Geometric Methods in Robotics. Springer, Berlin (1996)
187. Seube, N., Moitie, G., Leitman, B.: Aircraft take-off in wind shear: a viability approach. Set-Valued Anal. **8**, 163–180 (2000)
188. Serakos, D., Lin, C.F.: Three dimensional mid-course guidance state equations. In: American Control Conf., pp. 3738–3742 (1999)
189. Shaffer, P.J., Ross, I.M., Oppenheimer, M.W., Doman, D.B.: Fault tolerant optimal trajectory generator for reusable launch vehicles. J. Guid. Control Dyn. **30**, 1794–1802 (2007)
190. Shanmugavel, M., Tsourdos, A., Zbikowski, R., White, B.A., Rabbath, C.A., Lechevin, N.: A solution to simultaneous arrival of multiple UAV using Pythagorean hodograph curves. In: American Control Conference, Minneapolis, MN, pp. 2813–2818 (2006)

191. Shiller, Z., Lu, H.H.: Computation of path constrained time optimal motions with dynamic singularities. In: ASME Dynamic Systems, Measurement and Control, vol. 114, pp. 34–40 (1992)
192. Shin, K.G., McKay, N.D.: Minimum time control for robotic manipulators with geometric path constraints. IEEE Trans. Autom. Control **30**, 531–541 (1985)
193. Shue, S., Agarwal, R.: Design of automatic landing system using mixed $H2/H_\infty$ control. J. Guid. Control Dyn. **22**, 103–114 (1999)
194. Siciliano, B., Sciavicco, L., Villani, L., Oriolo, G.: Robotics, Modelling, Planning, and Control. Springer, Berlin (2009)
195. Singh, S.N., Steinberg, M.L., Page, A.B.: Nonlinear adaptive and sliding mode flight path control of FA 18 model. IEEE Trans. Aerosp. Electron. Syst. **39**, 1250–1262 (2003)
196. Solaque, L., Lacroix, S.: Airship control. In: Ollera, A., Maza, I. (eds.) Multiple Heterogeneous Unmanned Aerial Vehicles, pp. 147–188. Springer, Berlin (2007)
197. Solaque, L., Pinson, Z., Duque, M.: Nonlinear control of the airship cruise flight phase with dynamical decoupling. In: IEEE Electronics, Robotics and Automotive Mechanics Conf., pp. 472–477 (2008)
198. Sonneveldt, L., Van Oort, E.R., Chu, Q.P., Mulder, J.A.: Nonlinear adaptive trajectory control applied to an F16 model, J. Guid. Control Dyn. **32**, 25–39 (2009)
199. Soueres, P., Laumond, J.P.: Shortest paths synthesis for a car-like robot. IEEE Trans. Robot. **41**, 672–688 (1996)
200. Spooner, J.T., Maggiore, M., Ordonez, R., Passino, K.M.: Stable Adaptive Control and Estimation for Nonlinear Systems: Neural and Fuzzy Approximator Techniques. Wiley, New York (2002)
201. Stevens, B.L., Lewis, F.L.: Aircraft Control and Simulation, 2nd edn. Wiley, New York (2003)
202. Sunderhauf, N., Lange, S., Protzel, P.: Using the unscented Kalman filter in mono SLAM with inverse depth parametrization for autonomous airship control. In: IEEE Int. Workshop on Safety, Security and Rescue Robotics, vol. 1, pp. 1–6 (2007)
203. Sussmann, H.J.: A general theorem on local controllability. SIAM J. Control Optim. **25**, 158–195 (1987)
204. Sussmann, H.J.: Shortest 3-dimensional paths with a prescribed curvature bound. In: 34th IEEE Conf. on Decision and Control, pp. 3306–3312 (1995)
205. Takaya, T., Kawamura, H., Minagawa, Y., Yamamoto, M.: PID landing orbit motion controller for an indoor blimp robot. Artif. Life Robot. **11**, 227–234 (2007)
206. Thomasson, P.G.: Equations of motion of a vehicle in a moving fluid. J. Aircr. **37**, 630–639 (2000)
207. Turner, J.S.: Buoyancy Effects in Fluids. Cambridge University Press, Cambridge (1973)
208. Twaites, J.S.: Aerodynamics. Cambridge University Press, Cambridge (1960)
209. Ursem, R.K.: Models for evolutionary algorithms and their applications in system identification and control optimization. Ph.D. thesis, Univ. of Aarhus, Denmark (2003)
210. VanderBerg, J.P.: Path planning in Dynamic environments. Ph.D. thesis, Univ. of Utrecht, Netherlands (2007)
211. Van Der Zwaan, S., Bernardino, A., Santos-Victor, J.: Vision based station keeping and docking for an aerial blimp. In: IEEE/RSJ Int. Conf. on Intelligent Robots and Systems, Kagawa Univ., Takamatsu, Japan (2000)
212. Wang, X., Shan, X.: Airship attitude tracking system. Appl. Math. Mech. **27**, 919–926 (2006)
213. Watkins, A.S.: Vision based map building and trajectory planning to enable autonomous flight through urban environments. Ph.D. thesis, Univ. of Florida, Gainesville (2007)
214. Watt, G.D.: Estimates for the added mass of a multi-component deeply submerged vehicle. Technical memo 88/123, DREA, Canada (1988)
215. Wayshek, J., Dogan, A., Bestaoui, Y.: Investigation into the time varying mass effect on airship dynamics response. In: AIAA Aerospace Sciences Meeting, Orlando, FL (2009)
216. Wayshek, J., Dogan, A., Bestaoui, Y.: Investigation into the time varying mass effect on airship controller performance. In: AIAA Atmospheric Flight Mechanics, IL (2009)

217. Wayshek, J., Dogan, A., Bestaoui, Y.: Comprehensive characterization of airship response to wind and time varying mass. In: AIAA Atmospheric Flight Mechanics, Toronto, Canada (2010)
218. Wie, B.: Space Vehicle Dynamics and Control. AIAA Education Series (1998)
219. Williams, P.: Aircraft trajectory planning for terrain following incorporating actuator constraints. J. Aircr. **42**, 1358–1362 (2005)
220. Wilton, D., Rao, S., Glisson, A.: Potential integrals for uniform and linear source distributions on polygonal and polyhedral domains. IEEE Trans. Antennas Propag. **32**, 276–281 (1984)
221. Wolf, M.T., Blackmore, L., Kuwata, Y., Fathpour, N., Elfes, A., Newman, C.: Probabilistic motion planning of balloons in strong uncertain wind fields. In: IEEE Int. Conf. on Robotics and Automation, pp. 1123–1129 (2010)
222. Yakimenko, O.A.: Direct method for rapid prototyping of near optimal aircraft trajectory. J. Guid. Control Dyn. **23**, 865–875 (2000)
223. Yang, H., Zhao, Y.: Trajectory planning for autonomous aerospace vehicles amid known obstacles and conflicts. J. Guid. Control Dyn. **27**, 997–1008 (2004)
224. Zelikin, M.I., Borison, V.F.: Theory of Chattering Control. Birkhäuser, Basel (1994)
225. Zhang, H., Ostrowski, J.P.: Visual servoing with dynamics control of an unmanned blimp. In: IEEE Int. Conf. on Robotics and Automation, pp. 618–623 (1999)
226. Zhang, H., Ostrowski, J.P.: Periodic control for a blimp like dynamical robot. In: IEEE Int. Conf. on Robotics and Automation, Seoul, Korea, pp. 3396–3401 (2001)
227. Zhang, K., Han, Z., Song, B.: Flight performance analysis of hybrid airship. In: AIAA Aerospace Sciences Meeting, Orlando, FL, Paper AIAA2009-901 (2009)
228. Zhao, Y.J.: Extracting energy from downdraft to enhance endurance of uninhabited aerial vehicles. J. Guid. Control Dyn. **32**, 1124–1133 (2009)
229. Zhao, Y.J., Qi, Y.C.: Minimum fuel powered dynamic soaring of unmanned aerial vehicles utilizing wind gradients. Optim. Control Appl. Methods **25**, 211–233 (2004)
230. Zhou, K., Doyle, J.C.: Essentials of Robust Control. Prentice Hall, New York (1998)
231. Zipfel, P.H.: Modeling and Simulation of Aerospace Vehicle Dynamics. AIAA Education Series, 2nd edn. AIAA, Reston (2007)

Index

Y. Bestaoui Sebbane, *Lighter than Air Robots*,
Intelligent Systems, Control and Automation: Science and Engineering 58,
DOI 10.1007/978-94-007-2663-5,

MIX
Papier aus verantwortungsvollen Quellen
Paper from responsible sources
FSC® C105338

If you have any concerns about our products,
you can contact us on
ProductSafety@springernature.com

In case Publisher is established outside the EU,
the EU authorized representative is:
Springer Nature Customer Service Center GmbH
Europaplatz 3, 69115 Heidelberg, Germany

Printed by Libri Plureos GmbH
in Hamburg, Germany